AF784828

Contents

Zusammenfassung

Die Beiträge dieses Hefts untersuchen Anwendungen digitaler Technik und die Geschichte des Föderalismus. Themen sind das Verhältnis zwischen Staat und Gliedstaaten, sowie zwischen Staat, Information und Technik; Aushandlungen in der Migrations- und Hochschulpolitik; sowie die Koordination, die das Hochleistungsrechnen lokal und national erforderte.

Die Geschichte des Föderalismus entfaltet sich in der Digitalisierung vor dem komplexen Hintergrund von Träumen und Erwartungen, Kooperationen und Konflikten sowie von Kontinuität und Veränderung. Geleitet von einem gemeinsamen Ziel, waren individuelle und institutionelle Akteure gezwungen, sich mit einem ständig wechselnden Gleichgewicht von Ressourcen, Technologien und Verantwortlichkeiten auseinanderzusetzen.

Résumé

Les contributions de ce numéro se penchent sur les utilisations de la technologie numérique et sur l'histoire du fédéralisme. Les thématiques abordées sont la relation entre l'État et les États fédérés ainsi qu'entre l'État, l'information et la technologie; les négociations en matière de politique migratoire et d'enseignement supérieur; ainsi que la coordination requise aux niveaux local et national par le calcul à haute performance.

L'histoire du fédéralisme se déploie, dans le domaine de la numérisation, dans l'environnement complexe des rêves et des attentes, des coopérations et des conflits ainsi que de la continuité et du changement. Menés par un objectif commun, les acteurs individuels et institutionnels ont fait face à un équilibre en constante évolution des ressources, des technologies et des responsabilités.

Foreword

Like many ideas, the idea for this issue of *Itinera* came out of an informal meeting – a detail of history typically not preserved in archives. In 2017 two research teams from the Swiss Federal Institute for Technology (ETH Zurich) and the Università Svizzera Italiana Lugano (USI Lugano) met and decided to set up a joint research project. The inspiration for this decision was a historical source – a federal dispatch from 1985 announcing «special measures in favor of education and further training as well as research in information technology and engineering sciences».[1] This document led the two teams to collaborate on a project titled Digital Federalism, which had two objects of investigation – the SWITCH network and the supercomputing center in Manno, in the canton of Ticino, both of which are also addressed in papers in this issue.[2]

In an early stage of our research, we aimed to examine archival sources on the history of the network and of the supercomputing center through the lens of federalism. However, when the research started, we noticed that it took hardly any effort to trace the progress of federalism in the narratives provided by the archival materials we were perusing, because the many facets and nuances of the Swiss federal system shone through them all. We learned that chasing down specific developments – «automation», «computerization», and «digitization», for example – and including the perspectives of actors at different levels invariably involved federalism in a federal state. Quite soon, we realized that we were not alone in this journey. This special issue is the result of the dialogue we have woven over time with brilliant colleagues who helped us to look at the history of digital federalism from different theoretical and empirical points of view. Digital federalism and its imaginaries led us to the 2019 annual meeting of the Society for the History of Technology (SHOT) in Milan. In 2020 we discussed digital federalism from the point of view of the history of innovation with international scholars and members of the working group Geschichte und Informatik (G&I) in

1 Swiss Federal Gazette, 1986, 1: 321–383.

2 The research project «Digital Federalism. The Early History of SWITCH and CSCS Manno (1985–1995)» was funded by the Swiss National Science Foundation from fall 2018 to spring 2020, http://p3.snf.ch/Project-183007 (29/3/2021).

Bern. Sources analyzed in this issue of *Itinera* were jointly read and discussed at several meetings of the Computer, Administration, and History network in Weimar, Lüneburg, and Lübeck. We were furthermore lucky to profit from meetings and a conference in Lugano and from the Collegium Helveticum's academic workspaces in Zurich. Digital federalism was discussed on the occasion of several internal and public colloquia during the collegium's Digital Societies fellowship period (2016–2020).

Looking at history through the lens of digital federalism has been a great challenge for us as historians and social scientists. It requires both an individual and collective effort to bring different «eyes», sources, mindsets, and expertise into dialogue. Moreover, cooperation and dialogue were not just a feature of the histories covered in this issue, but also an essential element of our collective work as scholars.

As editors, we wish to acknowledge all the authors who accepted our invitation to contribute to this exciting project. We would also like to thank the Collegium Helveticum and the Chair for the History of Technology at ETH Zurich for financial support to publish this special issue. We are grateful to the editor of *Itinera* and to the General Secretariat of the *Schweizerische Gesellschaft für Geschichte* (*Société Suisse d'Histoire – Società Svizzera di Storia*) for integrating the issue into the series as well as for allowing us to publish it in English. We thank Giselle Weiss in the name of all the authors for the language proofreading and editing: her contribution to each and every paper was invaluable. We also gratefully acknowledge critical feedback (thank you, Silvia Berger Ziauddin and Ricky Wichum), anonymous reviewers, and the publisher, Schwabe Verlag, for editing, design, and printing. For extensive research on Swiss multilingual archival finding aids we thank Henrike Hoffmann. Finally, we owe a special debt of gratitude to all the archivists and technicians who so patiently managed to both satisfy and feed our historical curiosity. They protect and preserve all the precious traces we have followed so far – as well as those to be followed in the future.

Lübeck and Milan, summer 2021

Digital federalism: balancing automation, authority, and autonomy

Paolo Bory, Daniela Zetti

Federalism is about shared sovereignty. Like all social processes, it needs to be tested, revised, negotiated, and even changed according to the problem or decision at hand. In this issue of *Itinera* we argue that diverse and disparate processes of digitization have proved a good testing ground for federalism's many faces, as well as its strengths and weaknesses. By publishing a good handful of investigations in chronological sequence, we attempt to explain and understand federalism as a process, and thus as a historical subject.

For decades, digital technology has been seen equally as a powerful tool, a promise of progress unkept, and as «dumb» technology.[1] Digital technology requires coordination, as evidenced by the many terms typical for the field of computing that sound straightforwardly political. Indeed, sometimes these words – e.g., *protocol*, *center*, *network*, and *control* – do denote political and administrative matters. In Swiss federalism, implementing computing technology entailed the passing on of tasks to the next higher level according to the principles of subsidiarity and solidarity. Digital technology and automation of processes thus paved the way for coordination, but not in the sense of a basic menu of technology services. This is a general finding that emerged from the historical investigations behind the papers collected here. Whether the subject is the federal government's first computer, automated processing of punched cards, digital telecommunications, or supercomputers – digital

1 These observations are widely shared by studies on the history and philosophy of the computer as a universal machine – a machine that is conceptualized and materialized in diverse applications and narratives. See Michael Mahoney's seminal paper titled «The Histories of Computing(s)», first published in Interdisciplinary Science Reviews, 2005, 2: 119–135. Scholars emphasize the relevance of historical criticism in computing given that protagonists of computer development and digital technology produce their own models of historical development. Such models employ, for example, linear concepts of progress that are insufficiently complex for understanding historical change and social innovation. See, for instance, David Gugerli and Daniela Zetti, Computer History. The Pitfalls of Past Futures, Preprints zur Kulturgeschichte der Technik, 2019 (33).

technology and automation never provided coordination or balance per se; rather, they stimulated and even demanded it.

The automation and digitization of Swiss technology and society required substantial cooperation in terms of deploying and balancing political, economic, and human resources among the various actors involved.[2] As such, it posed practical challenges. But these same challenges also brought out the qualities and dynamism of federal systems. The development of digital technology opened up new fields of federal policymaking. In other words, while computing and networks put federalism to the test, federalism in turn exploited digital technologies as a demonstration of its qualities.

When reading the historical accounts of the protagonists who appear in this issue, readers may note that digital federalism has long been characterized by a constant need to balance three complementary elements: automation, authority, and autonomy. Indeed, in all the accounts collected in this issue, autonomy and authority are negotiated in order to reach a certain degree of automation within different social and material spaces. These spaces and places include administrative offices, scientific centers, and federal borders.

The fraught relationship between humans and machines is at the core of a discourse on automation that can be traced back to the 19th century's mechanization of work and today is found in the hopes and fears evolving

2 Historical studies on Swiss public and private projects of automation, computerization, and digitization have appeared since the early 21st century. Museum for Communication (ed.), Loading History. Computergeschichte(n) aus der Schweiz = Loading History. Chronique(s) de l'informatique en Suisse, Bern 2001. See also traverse 2009/3: Gesteuerte Gesellschaft / Orienter la société, www.revue-traverse.ch/ausgabe/2009/3/gesteuerte-gesellschaft (30/3/2021); several issues of Preprints zur Kulturgeschichte der Technik, https://www.tg.ethz.ch/produkte/preprints/ (30/3/2021) and of the book series Geschichte und Informatik – Histoire et Informatique, www.chronos-verlag.ch/reihen/2274 (30/3/2021). Peter Haber and Jan Hodel, Informatisierung / Informatisation / Informatizzazione, in: Historisches Lexikon der Schweiz, 2019, https://hls-dhs-dss.ch/de/articles/013724/2019-07-12/ (30/3/2021). For vivid accounts of the so-called liberalization of Swiss telecommunications, see www.oralhistory-pttarchiv.ch (20/8/2021).

around the social impact of the use of algorithms and artificial intelligence.[3] In the history of federalism since the 1950s, automation – both as a concept and a common goal – usually precedes the realization and spread of digital media and infrastructure. In most cases the (alleged) advantages of automated processes, whether promised by computers or digital networks, have an immediate impact on national, regional, and local policies even before the supporting technologies and infrastructure are in place. Words such as *optimization*, *simplification*, *modesty*, and *coordination* go hand in hand with the idea of automating human and organizational practices by means of technological innovation. Yet they don't automatically represent social or political innovation. This is another result of the research gathered here. The engineering projects analyzed in this issue succeeded in establishing digital federalism much more than in imposing rationalization or optimization.[4] In the 20th century, automation became a peculiar feature of federalism. It was repeatedly employed as a counterweight to achieve the desired balance between autonomy and authority.

Historical research has long since shown that federalism is no one-size-fits-all formula to be applied within a fixed regulatory political and cultural system. The *Historisches Lexikon der Schweiz* (*Dictionnaire Historique de la Suisse – Dizionario Storico della Svizzera*) states that «no single definition of

3 On the transformation of modern statehood and law facing the perils of industrial mechanization, see François Ewald, L'Etat providence, Paris 1986. Jon Agar studied how 19th-century experts attempted to control the UK administration thanks to the deployment of machines: The Government Machine: A Revolutionary History of the Computer, Baltimore 2003. For a classic study investigating the collective career of corporate IT personnel, see Thomas Haigh's Inventing Information Systems: The Systems Men and the Computer, 1950–1968, in: Business History Review, 2001, Special Issue 1: 15–61.

4 These results generally confirm older cultural studies on the relationship between society, science, and technology, which take into account the manifold relationships between human and non-human actors. In particular, the results affirm relational approaches used by humanity scholars in order to make digitized sociotechnical systems accessible to professional and regulatory actors, *and* for public discourse and civic participation. See, for example, Marc Coeckelbergh, Artificial Intelligence, Responsibility Attribution, and a Relational Justification of Explainability, in: Science and Engineering Ethics, 2020, 4: 2051–2068.

first time studies that reflect on digital federalism in the 20th century, this issue also represents a first attempt to put several cases in chronological order and to consider them by comparing and relating them to each other. Chronology and narration reveal the power of historiography to challenge what seems to be obvious. Individual and collective learning is based on reflection. It is reflection that makes information, like a historical work, valuable.

The first two papers in this issue deal with the relationship between and among states, information, and computers. The papers emphasize the long and varied history of information in state administration and the momentous consequences of technological visions. As early as the 1950s, Swiss federalism was addressed by governmental and administrative actors with the help of computer technology. Since then, implementing digital technology – in public sector administration for education and research purposes, in emerging sociopolitical and economic regions of Europe, and in global communities – implies activating and even innovating federal mechanisms that are anything but purely technical. All the steps along the way toward digital federalism – whether dreams or expectations, automation, cooperation, or conflict – are moreover embedded in a longer history that reflects a balance of forces, resources, and responsibilities that was constantly shifting among the various actors. The two articles in the middle of the issue investigate federal negotiations in the fields of migration and university policy. These articles emphasize how, in the second half of the 20th century, federalism was closely linked to the development of the welfare state, its economic and cultural resources, and its challenges. The last two papers deal with supercomputing, a prominent phenomenon of recent national and transnational history. The papers show how, by the end of the 20th century, the nation-state and

sult the following studies: Gregory Asmolov and Polina Kolozaridi, Run Runet Runaway: The Transformation of the Russian Internet as a Cultural-Historical Object, in: The Palgrave Handbook of Digital Russia Studies, Cham 2021; Paolo Bory, Gianluigi Negro, et al. (eds.), Computer Network Histories. Hidden Streams from the Internet Past, Zurich 2019; Gerard Goggin and Mark McLelland (eds.), The Routledge Companion to Global Internet Histories, New York 2017; Gianluigi Negro, Internet in China, London 2018; Valérie Schafer and Benjamin Thierry, Le Minitel: l'enfance numérique de la France, Paris 2012.

the very meaning of the computer were being challenged by the unprecedented «super» scale of the social and political coordination required by high-performance computing.

The first paper by Laura Skouvig shows that the relationship between state, information, and technology has been close for centuries. For this very reason, it is important to point out that the concept of the information state implies an ever-changing constellation of new technologies, different forms of information, and statehoods. Skouvig makes the case that the close nexus observable in digital societies is not new at all. Historiography has long maintained that different states had different reasons for collecting different forms of information in different ways. That is why her paper occupies the opening pages of this special issue. Based on an extensive body of research on the history of information, Skouvig argues that historiography has the power to challenge narratives that treat information as a naturalized entity. She shows that in the second half of the 20th century, a very popular concept arose that was rooted in communication theory and engineering and that saw information as quantifiable.

What is the quality of information? Why do states use technology? And why do they collect information? These questions are the most relevant because according to Skouvig current «[v]iews on quantity lead to determinist understanding of the information society», which evens out differences in attitudes toward information:

> Although the information state as an analytical concept has a tendency to focus on technologies in the form of death certificates, other printed forms, or the census, it raises questions about the underlying reasons and ideas for collecting information and for using particular technologies.

Information approaches ideology when neither civil servants nor learned scholars question its quality, to name just two categories of actors who have been custodians of information for centuries.

Nick Schwery's paper provides a first essential hint of the balance between authority and autonomy by approaching the arrival of the computer through the lens of information as well as of organization management in Switzerland. In reading Schwery's paper on the coordinating role of the Swiss Central Office for Organizational Affairs between the 1950s and the early 1960s, we noticed that more than being just a matter of content or data management during this period, information was first and foremost about organization. In fact, in the initial steps toward Swiss digital federalism, the computer appeared «on the horizon» as a key tool not only for managing the quantity of information and data, but especially as a technology adept at shaping the human, physical, and technical organization of the federal administration.

Schwery reflects on the many authorities involved in the automation process:

> [T]he computer became a vehicle for self-reflection, and subsequently a way to change the administration and the way it did its work under the radar of politics. The cooperation between highly heterogeneous institutions and actors stood at odds with the line organization.

As Schwery argues, the prospect of the computer changed the Confederation's administration, even before the actual arrival of the machines. Computing challenged the borders between the administration and the political sphere or, in other words, between autonomy and authority, opening up a whole new space of negotiation between different institutions and political actors. In Schwery's words, after a preliminary phase of coordination in the 1950s, by 1960 the negotiation ground for the future use of computers was defined «[n]ot in response to the computer, but anticipating computer use and computer access from a holistic standpoint. Not as an answer to the growth of the administration, *but in the context of it*» (emphasis added).

Once the computer entered the federal administration, the demand for automated decisions and standardized digital practices from Swiss institutions, offices, and organizations shot up. In this respect, Moritz Mähr's paper takes a step forward, both chronologically and pragmatically, toward attempts to implement federalism using digital means. Mähr takes a look at migration, one of the first core applications of computers for Swiss policymaking. In the 1960s the promise of automation informed the federal principle of distributing external human resources (but also burdens) throughout the country.

As Mähr points out:

> [B]etween 1967 and 1969 a solution was found which was also viable for the cantons. Not much remained of the visionary claim to automate migration policy. The information system was to replace the existing statistics without major adjustments in the cantons and municipalities. [...] The demand for radical control of migration flows had given way to the federal reality in which cantons and the federal administration negotiated a compromise.

The troubled path between 1964 and 1971 that led to the (partial) automation of the Central Aliens Register (ZAR) is a wonderful historical example of the constant negotiation between cantonal and federal institutions – in this case between the Federal Council, the Aliens Police, and the cantonal authorities. But Mähr's account does not dwell only on power relationships. It also shows how digitization challenged the heterogeneity of Swiss cantons in terms of work and organizational habits, social practices, material (and analog) tools, and methods of controlling migration. A clear example of the impact of automation on the material and sociotechnical dimensions of federalism is, in fact, analog technologies, such as the different paper forms designed and used by cantonal authorities on «the front line» – i.e., on national borders – which had to be standardized in order to be processed by a centralized computing facility. In the history of ZAR, these analog forms acted as brokering objects. They were the «material» of a complex negotiation between autonomy, automation, and authority, and as such between the federal administration and the Aliens Police or, in brief, between (federal) «offices» and (cantonal) «officers».

It is worth noting that although the first two case studies analyzed in this issue deal with different forms of sociotechnical networks, they do not involve digital networks. In other words, at an early stage of digital federalism, analog and digital means of communication and information processing coexisted. For example, in Mähr's account, migration forms and the statistical results output by computers were shared between cantonal and federal authorities by post.

The paper by Daniela Zetti is the first study in this issue to examine digital networks. It deals with the premises and special measures behind the scientific network known as SWITCH. At first glance, the essay seems to track a technological leap that took place between the early 1970s and the late 1980s, bringing readers into the age of digital telecommunications and networks. But most of Zetti's analysis refers to the late 1970s, shedding light on the negotiation of decision-making power and science policy between the Swiss universities and the federal state during this period. Zetti looks beyond the technological advancements embedded in the SWITCH network to the historical processes behind the federal dispatch of 1985 announcing «special measures in favor of education and further training as well as research in information technology and engineering sciences».[8] Contrary to other, enthusiastic accounts of national networking projects, and rather than focus on the «novelty» of the network – which, incidentally, was not new at all – Zetti chose instead to trace cooperative federalism in the lead-up to the federal «special measures» that gave rise to SWITCH. As Zetti puts it:

> The future SWITCH computer network and its supporting organization, the SWITCH Foundation, were innovative, and hence constituted special measures relating to university policy that relied on institutions and helped to establish rules and boards for discussion and decision-making. [...] With regard to the intended effects of the special measures – education, training, and managing structural change – the network most likely performed poorly or at least in a way that cannot be measured. [...] The positive societal impact is to be found somewhere else.

The «somewhere else» is again traceable in the political, economic, and cultural arena, where the federal state, universities, and computer scientists negotiated their respective claims on autonomy, authority, and automation. Zetti explores the potential of the dual nature of the network, in its «ideal and material form», looking at SWITCH as a point of departure rather than a terminus for historical investigation. This material and ideal form is discernable in several traits of the Swiss scientific and political system of the 1970s: the heterogeneity of the network «nodes», which represented different hardware and user practices at each university; the Swiss science institutions' claim of autonomy from the «special measures» that were eroding the universities' power to decide funding for educational and technological assets; and even the question of whether the upcoming networking projects

8 Swiss Federal Gazette.

were consistent with the principle of «frugality» peculiar to the Swiss federalist vision.

Returning to Laura Skouvig's plea for turning a historical gaze on «the quality of information», Zetti shows how the network was not just a simple technological solution for managing information. Rather, it constituted the enactment of the federalist hope and principles revolving around cooperation and planning efforts.

In line with this narrative, the last two contributions, which deal with supercomputing projects in Germany and Switzerland, respectively, probably represent the highest level of cooperation and balance between institutional and economic forces in digital federalism. Supercomputing is not discreet. In fact, it is conspicuous, state David Gugerli and Ricky Wichum at the outset of their paper. Supercomputing and nation-states have a special relationship, because supercomputing

> means the allocation of an extreme amount of resources in a single and very complex computing center. [...] Hence, supercomputing has always been a playground for powerful, splendid governments and their technoscientific programs.

Gugerli and Wichum proceed to show – contrary to this first and obvious relationship between supercomputing and central power – how a new kind of federalism emerged in the field of supercomputing. The subject of their study is the Center for High Performance Computing in Stuttgart, southern Germany. The early phases of German digital federalism in supercomputing were characterized by attempts to distribute resources and later by regional competitiveness. In a third phase, which began in the second half of the 1990s, a qualitatively new form of federalism emerged. Gugerli and Wichum speak of «a regime in which a broad variety of actors and programs participated» that could not be related to any «predefined model and no organizational standard for supercomputing». The new form continues to adhere to this characteristic: it is constructed in «trading zones» to create local configurations of supercomputing. And even here, there is no local *spiritus rector*, but rather «strong alliances, carefully designed forms of autonomy, and selective interrupts for the control of desired and disruptive interdependence.»

The final contribution is by Paolo Bory, Ely Lüthi, and Gabriele Balbi, who provide an impressive example of how balance can be created through negotiations among unequal partners. Their study deals with the founding of a national supercomputing center in Switzerland promoted by ETH Zurich and the canton of Ticino. Ticino had no university in the 1980s, but nevertheless it was chosen as a site for national supercomputing.

The Swiss National Supercomputing Center (CSCS) took shape in an advanced computing landscape at Swiss universities. Academic digital Switzerland already existed, brought with it its own traditions, was growing stably, and was well connected internationally. By including the installation of a supercomputer in the package of special measures, the federal government acquired the political authority to manage the new scientific field of supercomputing and to institutionalize it through a center. The federal government was able to do this because of the availability of scientific, administrative and technical expertise at the federal level. Thus, the federal government delegated supercomputing within the federal sphere to the ETH Board[9] and to one of its two federal institutes of technology, ETH Zurich. Both with respect to the digital-state technosphere, with its strong emphasis on higher education, and the routines of the federal organizations, continuities are evident in Swiss supercomputing.

The decision to bring supercomputing to the southern part of Switzerland promised to extend the Swiss digital landscape in a most spectacular way. But supercomputing strained the institutional fabric of federalism to a surprising degree. The organizational form of the center was immediately visible and became a surface on which to project exotic visions and cooperative schemes of unlimited scope. National supercomputing in Ticino demonstrated that federal politics was alive, but in need of grand visions, especially at the supra-regional and international level. Moreover, federalism requires solidarity, as the authors point out:

> Tensions aside, CSCS was also an act of «solidarity» with Ticino. Solidarity is a key principle in any federal association as it entails the faith (from the Latin *fede*) of those allies, whether states or cantons, who tend to support each other in order to strengthen (to make more *solid*) the cohesion and development of the entire nation.

9 The ETH Board is the supervisory body for Switzerland's two federal institutes of technology and four federal research institutes.

The article by Bory, Lüthi, and Balbi elucidates a fragment of contemporary history from the late 20th century because it shows how the cohesion between societies in the digital era is «handmade». Cohesion needs networkers. Much local work was needed to expand digital federalism, and many actors from a wide variety of social, economic, and political backgrounds were involved in constructing the center. Supercomputing in Ticino rapidly acquired fame beyond the region's borders. Accordingly, the supercomputing center demanded more solidarity from its management in faraway Zurich than initially anticipated. In the meantime, the center's location was a scandal of such proportions that, even at the time, Council of States member Giuseppe Buffi insisted its «history [...] must be written down».

A common thread links all the papers in this issue, namely, the co-development of a digital society and federalism in the second half of the 20th century relied on a common agreement to secure the balance of automation, authority, and autonomy. Such history shows how the prosperity of federalism depends on a continuous and complex sociotechnical investment in institutions, information, and infrastructure to keep up with the times.

Exploring ideologies of information

Laura Skouvig

Abstract

This article explores the intertwinement of state, information, and ideology, prompted by the question of why modern states consider more information and ever more sophisticated information infrastructures to be the ideal solution to every problem or crisis. These views are shaped by an ideology of information that originated in the immediate postwar period – an ideology that eradicates the longer history of information and instead stresses it as a quantitative phenomenon characterized by a constant newness. Yet a better understanding of the role of information calls for a critical history that focuses on, for example, the durability of the information state and how information has been understood and managed in previous societies. Such a historical focus will serve to strengthen critiques aimed at debunking current ideological constructions of information.

Introduction

Spring 2020. The coronavirus is spreading across continents and affecting billions of people. States respond differently and according to different governmental traditions. Many apply a range of digital technologies in their fight against the virus: facial recognition, data from credit cards, and forced or voluntary use of apps for tracking temperature and other health conditions.[1] Digital infrastructures – normally a little noticed backbone of society – suddenly become a central issue for everyone. Besides exposing the digital infrastructures, the coronavirus pandemic also amplifies the role of information in late-modern societies in two ways: 1) how information, misinformation, and disinformation mingle and produce a so-called infodemic;[2] and 2) the tight intertwinement between the state and information: tracking and tracing the trail of the virus relies on gathering massive amounts of information. The state needs to understand its own situation vis-à-vis the virus (how

1 https://www.berlingske.dk/internationalt/ny-app-sporer-coronavirus-har-du-delt-tog kupe-med-en-smittet (30/6/2020).

2 https://www.who.int/dg/speeches/detail/munich-security-conference (30/6/2020).

many people are ill? how does the virus spread? what is the hospital capacity?). At the same time, it needs to control and to direct that information (as well as populations).[3] In any case, the crisis exposes underlying assumptions about information and ideologies of information, comprising efficiency as well as problematic issues of quantity and superficiality.

Visions of a digitized, effective society constitute a strong undercurrent in digital politics. The increased use of digital infrastructures affects not only the relationship between state and citizen, leading to the construction of the digitally competent citizen and an entirely new understanding of digital citizenship.[4] It also transforms the traditional encounter between public authorities and citizens and has consequences for how information is understood and the role it plays in the interactions between public authorities and citizens.[5] However, information is not tied only to digital infrastructures, computer databases, and the digital itself. As Edward Higgs argues, the exchange, collection, storage, and use of information and data have been a central element for states for centuries. But the purposes for which information has been gathered, displayed, and stored have differed according to different types of states.[6]

This essay explores the intertwinement of state, information, and ideology prompted by the question of why states today think that more information and ever more sophisticated information infrastructures are ideal solutions to every problem or crisis.[7] Presumably, the idea of information's applicability partly stems from a specific ideological framing of information

---

3 Patrick Joyce, The State of Freedom: A Social History of the British State since 1800, Cambridge 2013, p. 59.

4 Arne Hintz et al., Digital Citizenship in a Datafied Society, Cambridge 2019.

5 Ida Lindgren et al., Close Encounters of the Digital Kind: A Research Agenda for the Digitalization of Public Services, in: Government Information Quarterly, 2019, 36: 437–436.

6 Edward Higgs, The Information State in England: The Central Collection of Information on Citizens since 1500, Houndmills 2004.

7 In understanding infrastructure, I rely on Geoffrey C. Bowker et al., Toward Information Infrastructure Studies: Ways of Knowing in a Networked Environment, in: Jeremy Hunsinger et al. (eds.), International Handbook of Internet Research, Dordrecht 2010.

as quantitative and always new. In discussing ideologies of information, I subscribe to Ronald Day's rather pragmatic understanding of ideology as «cultural, social, and political constructions of taste and action».[8] This understanding does not, however, make it less complicated to investigate and identify past ideologies. It is equally important to keep in mind the distinction between the political and discursive level and what sources reveal about everyday practices involving information.

Understanding the current intertwinements between state and information means questioning the particular ideology behind the diverse views of information as either disease or efficiency.[9] My claim is that both views depend on an ideology of information that eradicates its longer history and instead stresses information as a quantitative phenomenon characterized by constant newness. I begin by mapping contemporary 21st-century perceptions of information as defined by quantity. This line of reasoning also implies a particular construction of information in the past. I argue that these understandings are rooted in a conception of the history of information which – according to tradition – began in 1948 and emphasized information's newness. These understandings have led to a popular narrative that addresses the revolutionary transformations of the internet as a technology where information and communication are woven together.[10] I then show how the conceptualization of the information state presents a much broader view of the history of information. By invoking this broader view of past information cultures and exposing a variety of notions and ideological claims of information, I argue for a history of information that questions its inevitability and power in today's society.

8 Ronald Day, The Modern Invention of Information: Discourse, History, and Power, Carbondale 2001, p. 115.

9 My contribution builds on a presentation that I gave on 27 September 2019 in a workshop titled Digital Federalism in the History of Technology and Knowledge, 1970–1995, organized by Prof. David Gugerli and Dr. Daniela Zetti, both at the Collegium Helveticum. The argument is a slight paraphrase of Colin Koopman's main argument: to scrutinize information before communication – see Colin Koopman, Information before Information Theory: The Politics of Data beyond the Perspective of Communication, in: New Media and Society, 2019, 21: 1326–1343.

10 Ibid., p. 1327.

An era of information overload

One way of analyzing information as a quantitative phenomenon is through information overload as one of the prominent tropes of today's information society. A popular understanding of information overload connects it with the birth of the internet as a crucial threshold against which previous examples of information overloads and ways of handling them are measured and compared. However, as I show in this section, such measurements and comparisons tend to ignore different ideological conceptions of information. The risk of this approach is that current tendencies become impossible to describe, analyze, and understand because they lack a historical context.

The World Health Organization's (WHO) recent use of the term infodemic to describe the current flow of information suggests that we are perhaps on the brink of a significant new way of perceiving information. An infodemic is more than just an (over)abundance of information. The term evokes catastrophic images of a flood of unhealthy, unreliable information, of misinformation, disinformation, myths, and rumors that refers to the quality of information. But infodemic primarily implies a quantitative perception. As such it builds on the idea of information overload as a compelling and recurrent trope in today's ideological construction of information. The trope includes a view of information as simultaneously disease and cure. As Evgeny Morozov argues, new apps that organize and help people select and navigate the wild rivers of information reflect a «technological solutionism».[11]

For scholars seeking to compare present and past societies by putting numbers on the growth of information and data, sheer quantity is the only relevant measure in today's information ideology.[12] Quantity and technology are key identifiers of information as media scholar Mark Andrejevic writes:

> [T]he amount of mediated information – that which we self-consciously reflect upon as information presented to us in constructed and contrived formats (TV shows, movies, newspapers, Tweets, status updates, blogs, text messages, and so on)

11 Eugenia Siapera, Understanding New Media, London 2018, p. 263.

12 For one way of calculating the growth, see, for example, Viktor Mayer-Schönberger and Kenneth Cukier, Big Data: A Revolution That Will Transform How We Live, Work, and Think, New York 2014, pp. 8–10.

> via various devices including televisions, radios, computers, and so on – has surely increased dramatically, thanks in no small part to the proliferation of portable, networked, interactive devices.[13]

Andrejevic defines information as constructed by humans and mediated and dispersed by media. The focus on media in the definition facilitates a distinction between past and present in which information scarcity emerges as a dominant feature of past societies. However, the nature of media remains unclear, and this omission reproduces a general conception of individuals in the 17th century: How could they possibly experience an information overload given what would appear to be limited access to media?[14] Implicit in this conception is an understanding of media as printed mass media, which neglects the role of newsmongers and orally transmitted information. Andrejevic notes that a scarcity of media does not mean that no information was available. People have always had to process other types of information – just not of the mediated kind.[15] As Andrejevic points out, the amount of information typical for today stems from electronic devices: the more devices, the more information. According to Andrejevic, technology was a prominent factor in producing much more information, thus emphasizing an alleged increase in information quantity (and speed). This supposition evokes a particular construction of the past based on a pronounced difference in quantities of information between scarcity (the past) and abundance (the present), without considering whether information in the 17th century was the same as information in the 21st century.

13 Mark Andrejevic, Infoglut: How Too Much Information Is Changing the Way We Think and Know, New York 2013, p. 3.

14 Ibid., pp. 4, 9.

15 Ibid., p. 2. One might speculate whether Andrejevic's distinction between mediated and other kinds of information reflects the distinction between natural information and non-natural information. Any such distinction does, however, not make the assumption of increased quantity of information more valid. For a discussion of natural and non-natural information, see Sille Obelitz Søe, The Urge to Detect, the Need to Clarify: Gricean Perspectives on Information, Misinformation, and Disinformation, PhD thesis, University of Copenhagen 2016, https://komm.ku.dk/ansatte/?pure=da%2Fpublications%2Fthe-urge-to-detect-the-need-to-clarify(3ec2a23c-cac9-45de-8689-8a7e51b53998).html (15/11/2020).

Trivial counting of quantities of information and data becomes problematic when combined with equally trivial comparisons of technological revolutions. Such comparisons between the arrival of the internet with either the telegraph or, more notably, with the printing press often reflect a partial technological determinism because such comparisons tend to focus on the technology. In particular, comparisons with the printing press refer to the unprecedented amount of information that it churned out at the end of the 15th century and the beginning of the 16th century.[16] The stories of technological revolutions run the risk of neglecting the role of information and technology in the centuries between these technological leaps and even out differences in attitudes towards information. Information is neither universal nor naturalized. Hence, attention must be paid to differing perceptions of information in past societies that might not have focused on quantity or at least that might have understood it in distinctly different ways. Perhaps the 21st-century conception of information is unique in identifying quantity as a problem. Yet early modern scholars also reflected on an abundance of information. As Ann Blair has argued, far from regarding it as a problem, they enjoyed it with a kind of information lust. Blair's analysis shows how information overload is not a real quantity, but rather a perceived one. For early modern scholarship it became a crucial task to store, sort, select, and summarize information in order to preserve as much of it as possible.[17]

Blair's investigation of early modern information managing practices reveals that both the perception and understanding of information have changed over the centuries. In particular, quantity appears to have had a different meaning for learned scholars in early modern Europe. Consequently, it is tempting to wonder, as Frank Webster does, when and why quantity became the most influential denominator of today's information society to the extent that it led to qualitative changes.[18] Webster is highly critical of quantity of information as a decisive factor for modern society as opposed to

16 Mayer-Schönberger and Cukier, Big Data, p. 10.

17 Ann Blair, Too Much to Know: Managing Scholarly Information before the Modern Age, New Haven 2010.

18 Frank Webster, Theories of the Information Society, 4th ed., London 2014, p. 11. Webster is an English sociologist and author of the standard work on the information society.

previous social formations. Views on quantity lead to determinist understanding of the information society and a reductionist conceptualization of information as reified and itemized.[19] Quantity alone does not explain the prominent role of information in the modern world; but its importance emphasizes the need for exploring how information gained such momentum.

The information narrative

In this section I will discuss how information apparently emerged out of nowhere as a bright new phenomenon in the immediate postwar period. In many ways, the ideology of information as a quantitative phenomenon is tied to the development of communication technologies and computers in the postwar era. From the late 1940s to the 1960s, computers were basically advanced calculators capable of processing stunning amounts of information and doing it faster than previously. Their binary structure and their digital organization required a specific understanding of information as coded, binary numbers. Originating in the 1960s, the idea of networked computers (i.e., that computers could also transmit information) brought computers and communication theory together and resulted eventually in the common conception of information as bits.[20]

Many information historians seek to understand how information became a central element in the ideological construction of the information society as a social formation from the 1960s to the beginning of the 21st century. Webster argues that this entanglement calls for a critical investigation in the form of a «Foucauldian account of the genealogy of ‹information›».[21]

A genealogical approach focuses on how the construction of the past is a crucial element in the ideology of information. Such an approach emphasizes

19 Frank Webster, The Information Society Revisited, in: Leah A. Lievrouw, Sonia Livingstone (eds.), The Handbook of New Media: Social Shaping and Social Consequences of ICTs, London 2006, pp. 449–451.

20 Paul. E. Ceruzzi, Computing. A Concise History, Cambridge MA 2012, pp. 4–12. See also Ronald R. Kline, The Cybernetics Moment: Or Why We Call Our Age the Information Age, Baltimore 2015, p. 6.

21 Webster, Revisited, p. 452.

critical questioning of the platform from which scholars today present, discuss, and criticize information ideology.[22] A genealogical approach thus addresses the blind spots of current information ideology, and also invites investigation of the past by looking at everyday routines.[23] Though I do not apply a fully genealogical approach in this essay, I share the commitment of such an approach: to elucidate the past's much more complicated dealings with information than the question of quantity suggests.

The critical promise of a genealogical approach is apparent in the identification of a consensus of tradition which argues that contemporary understandings of information were fully established in 1948 with the publication of Norbert Wiener's and Claude Shannon's individual seminal works and the overall acceptance of so-called information theory in the following decades.[24] As the American philosopher Colin Koopman argues, the tradition states that information gained an entirely new meaning through the work of Wiener and Shannon, as well as of Warren Weaver. Yet, asserts Koopman, this meaning was actually not new but was already culturally accepted by the end of the war.[25] The strength of this tradition can be seen in how quickly Shannon's theory was renamed information theory, though he devised it as a theory of communication having to do primarily with information carriers and very little with information in and of itself.

In the following, I briefly recapitulate the postwar narrative of the birth of information and how it became identified with information theory. Claude Shannon was an American engineer who worked for Bell Labs. Shortly after the Second World War, he devised what he called the mathematical theory of communication.[26] The aim of this theory was to solve the problem

22 For a detailed argument for choosing a genealogical approach see, for example, Colin Koopman, How We Became Our Data: A Genealogy of the Informational Person, Chicago 2019, pp. 22–24.

23 Michel Foucault, Nietzsche, Genealogy, History, in: Paul Rabinow (ed.), The Foucault Reader, New York 1984, pp. 76–101, on p. 76.

24 Koopman, Information Theory, p. 1328; see also Kline, Cybernetics Moment, pp. 102–135.

25 Koopman, Our Data, p. 182.

26 Claude Shannon, A Mathematical Theory of Communication, in: The Bell System Technical Journal, 1948, 27: 379–423, 623–656. Reprinted in Claude E. Shannon and

of sending messages undistorted through a communication system, which implied a particular definition of information. Shannon's main interest was communication and information related to the uncertainty of information selection – not the actual message to be communicated.[27] In Shannon's work information was a purely quantifiable phenomenon and the inherent definition of information was devoid of semantics and meaning.[28] However, as Robert Kline shows, this inherently mathematical theory of communication gained momentum and, in combination with Wiener's different and at the same time closely related cybernetics, became the core of the increasingly popular and popularized information theory – as a naturalized narrative.[29] Combined with Wiener's definition of information as a function of communication, information theory made its way into the social world beyond its origin in engineering.[30]

Whereas this shift in labelling might not be new to historians of information or scholars of information studies, the annexation of Shannon's theory by a wider scientific community and its renaming were important elements in the hyperbole of information in the years that followed. According to prominent information studies scholars such as Ronald Day and Geoff Nunberg, the consequence of the naturalized narrative was that it made it almost impossible to question the ideological formations of information and information theory.[31] Moreover, the narrative created the foundation for a common discourse that defined digital media by their newness.[32]

Warren Weaver, The Mathematical Theory of Communication, Urbana 1964, pp. 29–125.

27 Kline, Cybernetics Moment, p. 16.

28 Shannon, Mathematical Theory, pp. 31–32.

29 Kline, Cybernetics Moment, p. 6. Under the name of *information theory*, Shannon's brief conceptualization of information had a huge impact on how information was understood in, for example, information science.

30 Ibid., p. 123 ff. One step in this process was an article published by Warren Weaver that introduced Shannon's work to a wider audience in 1949. See Shannon and Weaver, Mathematical Theory.

31 Day, Modern Invention, and Geoffrey Nunberg, Farewell to the Information Age, in: Geoffrey Nunberg (ed.), The Future of the Book, Berkeley 1996, pp. 103–138.

32 Koopman, Information Theory, p. 1328.

federalism is possible».[5] The contributions in this issue show that autonomy and authority were politically and socially relevant throughout the period under investigation. Moreover, the two concepts characterize digital federalism's complex practices of negotiation. Autonomy and authority are closely related to each other. On the one hand, autonomy demands authority at the local and regional level. An example is when cantonal governments or universities make economic or political decisions based on their right to choose what is best for their own sociotechnical and cultural environment.[6] Conversely, in some cases authority bypasses local autonomy. Such is the case, for instance, when the federal government makes decisions on «overarching» projects that concern the country as a whole, with the consequent deployment of human and economic resources that no autonomous regional or local actor can afford.

Federalism is not synonymous with the nation-state. It can be practiced within and across states. The collection in this issue thus supplements accounts of the history of digital technologies and media from national, international, and transnational perspectives that have arisen within fields such as media studies, Internet studies, the history and philosophy of technology, computer history, and science and technology studies.[7] By collecting for the

5 Rainer J. Schweizer and Ulrich Zelger, Föderalismus / fédéralisme / federalismo, in: Historisches Lexikon der Schweiz, 2019, https://hls-dhs-dss.ch/de/articles/046249/2009-11-05/ (30/3/2021). For a concise history of federalism, see Dieter Langewiesche's take on «another German history». Langewiesche traces the history of German federalism from medieval times up to the present and «from a multi-state empire to a federal state». Dieter Langewiesche, Vom vielstaatlichen Reich zum föderativen Bundesstaat. Eine andere deutsche Geschichte, Stuttgart 2020. For an overview of the often, yet not exclusively national connotations of the term *digital sovereignty*, see Stéphane Couture and Sophie Toupin, What Does the Notion of «Sovereignty» Mean When Referring to the Digital?, in: New Media & Society, 2019, 10: 2305–2322.

6 The federal state has its own autonomy too. For a comparison of Swiss and German federalism, see Dietmar Braun, Dezentraler und unitarischer Föderalismus. Die Schweiz und Deutschland im Vergleich, in: Swiss Political Science Review, 2003, 9: 57–89.

7 Much of this research has been published as monographs or in academic journals, usually focusing on single case studies covering a short time span. An exhaustive list of all the contributions dealing with national case studies on digitization would be too long. However, in addition to several efforts from the United States, readers might wish to con-

Kline investigates how the information society and information technologies became central tropes in what he identifies as the techno-revolutionary narrative of 21st-century society.[33] As Kline defines it, the naturalized narrative consisted of a tight intertwinement between communication, technology, and information, which resulted in information becoming an inseparable element of communication. Information was even presupposed in communication, making it untouchable from a communication theory perspective.[34] Information gained status as a «reified token» that information professionals and theorists only rarely questioned critically. For those who did engage in such questioning, the «‹pregiven› character» of information hampered proper investigation.[35] It is just such a naturalized reified social construction that genealogy as a theoretical approach aims to debunk by repositioning it, as Koopman suggests, chronologically (before 1948). He further adds an epistemological repositioning by emphasizing the «in-for-mat-ting» aspects of information prior to any communication.[36]

An important assumption underlying Koopman's approach is that neither Shannon nor Wiener ever invented an entirely new definition of information unconnected to common language usage.[37] Koopman's assumption corresponds well with the findings of John Durham Peters, who in an article from 1988 on the history of information gave a general outline of the different meanings and uses of information from its first appearances in the English language in the Middle Ages until the Second World War.[38] Peters argued that one of the many changes stemmed from the incorporation of information into bureaucratic usage with reference to statistics.[39] According to Koopman, Shannon and Wiener did not so much contribute to the forma-

33 Kline, Cybernetics Moment, p. 203.

34 Koopman, Information Theory, p. 1332.

35 Day, Modern Invention, p. 115; Nunberg, Farewell, p. 107.

36 Koopman, Information Theory, p. 1334.

37 Koopman, Our Data, p. 182.

38 John Durham Peters, Information: Notes toward a Critical History, in: Journal of Communication Inquiry, 1988, 12: 9–23. Colin Koopman explicitly refers to Peters' analysis as justification for his own view.

39 Ibid., pp. 14–15.

tion of information as merely presuppose it for communicative means.[40] In other words, Shannon and Wiener relied on an existing understanding of information as quantifiable and universalizable that made it possible to transfer information to the emerging problem of communication.[41]

Information's status as pregiven and presupposed in communication theories across the technology and social sciences lays emphasis on another feature of information: it ostensibly lacks a history.[42] Information has a particular relationship with temporality that can be characterized by newness and its connection to the present moment: new information makes existing information obsolete and irrelevant.[43] New information tends to wipe out the past, Peters assumes, due to its inherent connection with science where results only stand until they are surpassed by new results.[44] As newness, information marks a difference, a short range of time that makes a difference. It is almost like a switch. Information possesses two important features for becoming ideologically important: first, it was (and still is) a word that is used in everyday life; second, as a word without history it is open to present and future ideas and utopian imagination.[45]

Peters argues that the success of information theory resulted from making a familiar experience from everyday bureaucratic meetings «into a lofty concept of science and technology».[46] The humble and very mathematical theory of communication became popularized as a general theory for understanding human communication. In that form, information became connect-

---

40 Koopman, Our Data, p. 183. Shannon was definitely cautious about the differences between what he saw as communication theory in a narrow sense (for engineering purposes) and information theory (the popularized and expansive version that Wiener embraced). However, Kline convincingly argues that Shannon did use information theory in his own writings, and his boundary work did try to restrain and yet also expand information theory. Kline, Cybernetics Moment, p. 103.

41 Koopman, Our Data, p. 182.

42 Ibid., p. ix. See also Peters, Information, p. 10.

43 Peters, Information, p. 19.

44 Ibid., p. 20.

45 Ibid., p. 17; Day, Modern Invention, p. 117.

46 Peters, Information, p. 18.

ed with visions of utopia and it came to encapsulate the idea of the modern society.[47] Information became ideologized:

> «Information» is a central term of ideology because it determines and patrols its own meaning over a vast expanse of social and cultural spaces. Through information, vocabularies for the future are included or excluded, shaping history in a way that is fit for information and for little else.[48]

Day not only discusses how Weaver's and Wiener's popularization of information theory as a general theory encapsulated utopian promises. He also argues for how early European documentation scientists such as Paul Otlet and Susanne Briet contributed to and were part of connecting information with utopian visions of a better and peaceful world based on efficient systems of knowledge organization and information management.[49] Day's analysis of documentation science provides a welcome perspective on the scientific enlistment of information before 1948. Kline's and Day's analyses of how information became interwoven with technocratic visions of a more efficient society during the 20th century illuminate a central condition in the futuristic vision of information: that the industrial society in and of itself could not fulfill the utopian promises of an information revolution which paved the way for the information society and its ideology of information highways.[50] Information as an ideological construction would thus seem to be a child of the 20th century based on quantity and technology as central components. Big data melds these two together.

Unraveling histories of information

From my position within information history, I share Koopman's insistence on investigating information before the 1948 crystallization of information theory. But the question is how to scrutinize a totally ideology-infected concept that seems impossible to investigate. Or, as Koopman asks: What was

47 Kline, Cybernetics Moment, p. 203.

48 Day, Modern Invention, p. 117.

49 Ibid., pp. 7–38.

50 Webster, Revisited, p. 445.

information before information theory?[51] In his own analysis he turns to the beginning of the 20th century, where he identifies a shift in information – from being universal to becoming universalizable.[52] In other words, information is not only «already everywhere» (universal) but now also «can be mobilized to operate anywhere we want it to»[53] (universalizable). The distinction has an important role in Koopman's argument, because it facilitates a demarcation of information technologies of previous centuries. As «antecedents» or «prototypes» they missed the scalability of universalizable technologies and represent not yet stabilized technologies.[54]

However unintended, Koopman's differentiation between information as universal or as universalizable runs the risk of predating the birth of information – now at the beginning of the 20th century. This leads to a crucial set of questions: How far back can we trace the history of information? And with what understanding, definition, or conceptualization of it? Inherent in the formulation of these questions seems to be the acknowledgement of a change in the meaning of information: a historical meaning and a «new» meaning. Koopman's definition of the change regards the claim of «universality» in information, and he positions it at the beginning of the 20th century. Other historians, such as Neil Postman and Toni Weller, argue for a mid-19th-century intersection between an early modern and a modern conception of information partly brought about by the telegraph but also by cultural notions.[55] They both argue that information became abstracted in this period as morsels of communication. Robert Darnton, however, argues for information as snippets in 18th-century Paris.[56] Apparently, we can keep pushing the threshold for a modern understanding of information back in time. As Geoff Nunberg argues, being abstracted seems to be the most prominent character-

51 Koopman, Our Data, p. 17

52 Ibid., p. 9.

53 Ibid., p. 10.

54 Ibid., p. 28.

55 Neil Postman, Building a Bridge to the 18th Century: How the Past Can Improve Our Future, New York 1999, and Toni Weller, The Victorians and Information: A Social and Cultural History, Saarbrucken 2009.

56 Robert Darnton, Poetry and the Police: Communication Networks in Eighteenth Century Paris, Cambridge MA 2010.

istic of the 20th-century definition of information. But he also points to etymological explorations of information indicating that this «abstractedness» has been inherent in information from its first appearances in medieval English.[57] In early modern Venice, *communicazione* signified how information was passed from one governing council to another.[58] Information is already always there in the shape of the presupposed content of communication. Koopman's distinction draws needed attention to a discussion of whether information suddenly gained a new meaning and also when (if ever) this shift took place.

Two points are worth keeping in mind in discussing the birth of information. First, the history of information is not just about dating when information first appeared in its modern meaning. Second, as both Nunberg and Paul Duguid emphasize, historians need to be highly sensitive towards the risk of presentism in locating 21st-century understandings of information in historical periods where information had entirely different meanings.[59] Perhaps a history of information could benefit from another aspect of Michel Foucault's genealogical approach that is attentive to emergence rather than origin.[60] What is interesting within a Foucauldian line of reasoning is not the true origin (birth) of information but emergences and transformations of information practices.[61] It remains crucial to investigate information before it became naturalized as information theory. What is at stake here is addressing the given historical, cultural, and social contexts of information – not determining it in relation to its present meanings. This is in line with Darnton's suggestion to regard all ages as information ages, but then to carefully scrutinize shifts and ruptures.[62] Weller argues for a longer horizon when

57 Nunberg, Farewell, p. 110.

58 Filippo de Vivo, Information and Communication in Venice: Rethinking Early Modern Politics, Oxford 2007, p. 4.

59 Nunberg, Farewell, p. 110, and Paul Duguid, The Ageing of Information: From Particular to Particulate, in: Journal of the History of Ideas, 2015, 76: 347–368.

60 Foucault, Nietzsche, p. 80 ff.

61 Ibid.

62 Robert Darnton, 5 Myths about the «Information Age», in: Chronicle of Higher Education, 2011, 17 April, http://chronicle.com/article/5-Myths-About-the-Information/127105/ (15/11/2020).

studying information historically, because «notions of what constitutes information [...] have not remained constant over time».[63] Indeed, she maintains, a history of information should explore exactly such changing definitions.

Most prominent is Weller's interest in wringing information from technology's firm hold on it in favor of investigating it as a cultural and social phenomenon.[64] I suggested at the outset that the (information) state is a nexus for historical inquiry into information due to the state's present proximity to information. However, anchoring information within the state might not minimize the role of technology. «The government machine», as Jon Agar frames it, signals a close relationship between state, information, and technology. This closeness should not shade the importance of the impact of and the dualistic relationship between cultural perceptions of information and the state's implementation of different information technologies.[65]

Within the frames of state bureaucracy and the state's need to know, communication infrastructures and information mingled.[66] In Patrick Joyce's view, a state is always dependent on its lines of communication, and the relationship between the state and its communications systems reveals much about the nature of the state.[67] Agar adds to this by investigating the most prominent of all state metaphors: that of the machine. He ties the considerable standardization and routinization of British bureaucracy to the development of Babbage's analytical engine, Turing's universal machine, and finally the computer. Their similarities, he claims, are to be found at another level, «because they were imagined in a world in which a particular bureaucratic form – an arrangement of government – was profoundly embedded».[68]

63 Toni Weller, Information History – an Introduction: Exploring an Emergent Field, Oxford 2008, p. 18.

64 Ibid., pp. 11–22.

65 Jon Agar, The Government Machine: A Revolutionary History of the Computer, Cambridge MA 2003.

66 Joyce, State of Freedom.

67 Ibid., p. 20.

68 Agar, Government Machine, p. 69.

An ideology of weaponry

Going beyond 1948 and even beyond the 20th century opens a vast field of ideas and meanings about information. These included perceptions of quantity, technology, and temporality, but also control of information and anxiety about losing it. As Postman points out, information was invisible in 18th-century parlance, yet he argues that it was ideologically important and used as such by, for example, the French Encyclopedists.[69]

A multiplicity of ideologies and cultures of information was prevalent in early modern Europe. Cultural differences could be defined by different geographical borders or by intellectual aspirations. The English form of government had a totally different basis compared with continental absolutist regimes like those of the French and Danish. And even between these two examples of absolutism there were differences. Ann Blair and, to some extent, Jacob Soll point to how information gathering was defined by a desire for information and the urge to collect and compile as much of it as possible. Blair looks at scholars in the Renaissance period, whereas Soll has investigated the information system of French minister Jean-Baptiste Colbert during the reign of Louis XIV in 17th-century France.[70] The urge to know resulted in practices of managing, storing, and organizing massive amounts of information. Soll argues that Colbert's lust for information stemmed from an understanding of information and knowledge as a means of securing power. The ideology of the absolute state held that power was secured in the hands of the sovereign and confirmed by divine right. Accordingly, the sovereign also had the right to know and to decide what information «was» and to whom it should be dispersed. It is obvious that a pronounced ideological foundation would spill over into the understanding of information and the information infrastructures between state and society as strongly hierarchical and vertical in structure.[71] This assumption, however, raises the question of how infor-

69 Postman, Building a Bridge, p. 86.

70 Blair, Too Much; Jacob Soll, The Information Master: Jean-Baptiste Colbert's Secret State Intelligence System, Ann Arbor 2011.

71 Ellen Krefting et al., En pokkers skrivesyge. 1700-tallets dansk-norske tidsskrifter mellom sensur og ytringsfrihed, Oslo 2014.

mation was understood and how it was practiced, for example, in systems for managing information.

In order to approach information historically, the state offers huge potential for examining material manifestations of information.[72] Such material manifestations (e.g., in petitions, files, and ledgers) can also lead to exploring the views, conceptualizations, and ideologies of information in past societies. The challenge when referring to the state and its ties with information and technology lies in the state's long endurance and history. Throughout this long history, the state in very different forms has had a continuous tradition of collecting, storing, sorting, and managing information using a variety of technologies. Registers of people moving to or leaving parishes or districts in Copenhagen as a form of control over vagrants and strangers were, together with censuses, key examples of a state's need for numbers.[73] Implementing new technologies such as the steel pen as well as bureaucracy reflected a new need for processing and structuring information in ways that were considered more appropriate if not more efficient.[74] From the perspective of a *long durée*, information and technology might take on exactly this character of being pregiven, already always there, and with a universalistic element, which Paul Duguid warns against.[75] Being attentive to the specific cultural and social contexts remains crucial for all historians.

The history of the state normally pivots around concepts like modernity and the nation-state. The English historian Edward Higgs suggests that the notion of the information state might serve to bridge the divide between early modern and modern (nation)-states. The latter, in the terminology of Anthony Giddens, points to the major rupture that occurred in state organization in the 19th century. Giddens's argument is that the modern nation-state was characterized by a massive need for information that, combined with surveillance and capitalism, marked it as distinctly different from previous

72 Agar, Government Machine, p. 3.

73 Grethe Ilsøe, Den enkelte og forvaltningen. Registrering som parameter, in: Karl Peder Pedersen, Grethe Ilsøe, Ditlev Tamm (eds.), På given foranledning. En antologi om dansk forvaltningskultur. Copenhagen, 1994, pp. 149–168.

74 Joyce, State of Freedom.

75 Duguid, Ageing of Information, p. 348.

state forms and that relied on the state's centralized administration.[76] Higgs investigates the English state between 1500 and 2000 using the conceptualization of the information state. From this perspective, he points to how early modern states gathered information that was decentralized in accordance with their structure. The decentralized information-gathering strategy is no less sophisticated than the centralized efforts of the modern nation-state, as Giddens suggests. Joyce furthers Higgs's inherent critique of Giddens by pointing out that the local state was as much «the state» as the central state was.[77] The delegation of authority from the central to the local state relied, however, on surveillance by the central state and on control using information and intelligence.[78] Although the information state as an analytical concept has a tendency to focus on technologies in the form of death certificates, other printed forms, or the census, it raises questions about the underlying reasons and ideas for collecting information and for using particular technologies. The inherent critique that the early modern state lacked solid information technologies probably lies in the deduction that few archived records exist. The impact of communication technologies on society and state formations has been debated since Harold Innis and the Toronto School. The British anthropologist Jack Goody has also pointed out how the logic of writing formed bureaucracies and, eventually, states.[79] Written archival records do, however, reflect a certain understanding of information transactions. Oral oaths were legally just as conclusive as a signed document, and many disputes were negotiated and concluded orally. Even in oral situations, testimonies were needed and, as such, information was increasingly used as a basis for decisions. Information needed to be pegged and stabilized in order to be controlled.[80] The state, according to Joyce, emerged out of writing.[81] The need to know manifested itself in a need to secure the knowledge.

---

76 Anthony Giddens, A Contemporary Critique of Historical Materialism, 2 vols., Vol. 2: The Nation-State and Violence, Cambridge 1985.

77 Joyce, State of Freedom, p. 25.

78 Ibid., p. 25.

79 Jack Goody, The Logic of Writing and the Organization of Society, Cambridge 1986.

80 Laura Skouvig, The Raw and the Cooked, in: Johan Östling et al. (eds.), Forms of Knowledge: Developing the History of Knowledge, Lund 2020, pp. 107–123.

81 Joyce, State of Freedom, p. 78.

A central component of Colbert's information system was that all knowledge had practical value for politics. As a minister at the height of absolutism where royal power was discursively legitimized through religion, no further legitimization appeared necessary. Colbert, however, saw a stronger need for substantiating royal power with reference to historical documentation and evidence, making state maintenance into a craft for archivists and traveling informants.[82] Yet, in line with absolutism, Colbert's information system was highly centralized and also personalized, which meant that it basically did not survive him. Colbert seemed to be the embodiment of the «I am the state» trope in French absolutism, or at least his information system became a junction in statecraft, connecting policy with information-handling practices.[83] The state was in favor of technologies that helped it to gain information and relied on different kinds of communication technologies and infrastructures. In Joyce's examination of the British state in the 19th century, the pivot is the materiality of technology as well as the actual places – e.g., the local post office – that helped naturalize the state.[84]

Colbert was not of a scholarly mind, but as a high-ranking civil servant he shared the same interest in organizing and controlling information that, for example, learned scholars had. Control of information and the maintenance of secrecy were absolutely core issues in early modern state practices, though not necessarily addressed explicitly. The desire to control information is a striking difference between states and scholars as representatives of the rising public sphere. Particularly the state represented a need for securing information and keeping it secret from a wider public. This wish crossed cultural, temporal, and geographical lines, as Filippo de Vivo shows in his analysis of Venetian information practices in the 17th century – a practice that is also detectable in Danish absolutist information ideologies. Information was dangerous because it could incite the population to revolution, and thus needed to be controlled in order to control the population. Means of control involved limiting printed news media such as newspapers but also, for exam-

82 Soll, Information Master, pp. 140–153.

83 Ibid. It is disputed whether Louis XIV ever said «L'État, c'est moi». It has, however, been commonly accepted as a way of condensing absolutism.

84 Joyce, State of Freedom, p. 53 ff.

ple, distributing networks of ballad-mongers.[85] Gathering information was meant to prevent certain groups from accessing the information.[86] Information itself became a way of securing control. As de Vivo argues, communication – and consequently information – marked the limits of the state's authority. States did not gather only dangerous information but also information about those who spread it. The population could only become visible to the state as information, ultimately entangling the population in a net of writing that made them legible to the state.[87]

For the scholars of the Renaissance, the heavy loss of manuscripts during the early Middle Ages had a profound impact on their attitude towards information.[88] One way of protecting information was to share it with others. Openness, access, and transparency can be seen behind the explosion of reference books, compilations, and encyclopedias. The French Encyclopédie is a prime example of openness and access to information. Postman interprets the Encyclopédie as a manifestation of the 18th century as an information age and as a model for the universalistic endeavor. As an information technology, its universal aspirations were only relevant in a certain context, defined by a specific rhetorical purpose, and giving shape to a concept that was about empowering and facilitating skepticism and critique of the existing system.[89] Information was a weapon against the absolute state. Secrecy, as opposed to transparency, could be seen as a demarcation line between scholars and states in ideological perceptions of information; yet secrecy was also prevalent for guilds and fraternities, for example, not to mention associations such as the Freemasons.[90] For Colbert, it was a strong impetus for gaining control of information and of those who had access to information at least in the

85 Laura Skouvig, Records and Rumors: Surveillance and Information in Late Absolutist Denmark (1770–1849), in: Surveillance & Society, 2017, 15: 314–325.

86 De Vivo, Information and Communication, p. 12 ff.

87 James C. Scott, Seeing Like a State: How Certain Schemes to Improve the Human Condition Have Failed, New Haven 1999, p. 2; Koopman, Our Data, p. 4 f.

88 Blair, Too Much, p. 22.

89 Postman, Building a Bridge, p. 86.

90 Elain Leong, Alisha Rankin (eds.), Secrets and Knowledge in Medicine and Science 1500–1800, London 2011; Reinhart Koselleck, Kritik und Krise. Eine Studie zur Pathogenese der bürgerlichen Welt, Freiberg (1959/1973) 2013, pp. 49–81.

form of «secrets of the state». In this respect, he needed to obtain control of parliament, and for this purpose he incorporated the Royal Library into his information management system. The Royal Library changed from aspiring to universalism to becoming a state information depot with selected papers, books, and other kinds of documents that Colbert considered to have political use.[91] Koopman points to the insufficiency of 18th- and 19th-century technologies in achieving their universalistic aims when implemented in practice. But this might be too simplistic an interpretation in light of Colbert's skilled exploitation of existing and new information management techniques.

To a large extent, however, the relationship or even opposition between state (information control) and the public sphere (transparency) neglects the influence of government agencies on the flow of information and also idealizes the public sphere. In the 16th and 17th centuries, the Venetian government was obsessed with secrecy – on a discursive level – because it proved difficult to conduct it in real life. Information was a tool and even a weapon in struggles over government strategies between secrecy and transparency.[92]

Conclusion: information as a recurring problem

Many scholars identify the period immediately after the Second World War as a central moment in the history of information. A certain ideological construction of information emerged that came to define the information society as it evolved from the 1960s onward.[93] This ideological construction not only defines information in a specific way. It also encapsulates a particular understanding of the role of history in the construction of information.[94] Although the hyped ideologies of information in the 1960s that focused on its utopian promises seem for now to have been replaced by a more dystopian worry about infodemics and the misinformation and disinformation they spread, information remains a highly ideological term. The naturalized narrative of

91 Soll, Information Master, pp. 95–97.

92 De Vivo, Information and Communication, p. 4.

93 Koopman, Information Theory; Kline, Cybernetics Moment.

94 See also Day, Modern Invention, and Peters, Information.

the birth of information, as a word without history, in the postwar period is reflected in, for example, WHO's use of the terms infodemics, misinformation, and disinformation.[95] Once again, in the present information society, information has become a problem caused by technology.

This brief look into early modern information cultures shows that information was prevalent and appeared in unprecedented formats and quantities. States and scholars alike produced more information (for different purposes) and sought to work out systems for controlling, managing, and disseminating information. The urge to keep up with information resulted in sophisticated tools and classifications that prevail to this day and has kept information stabilized and fixed. That we to some extent still rely on tools and managing systems developed in the period from the 17th century to the early 20th century is not, however, the way past information strategies are reflected in present ideological constructions of information and the information society. The past is used as an element in the construction of present information ideologies with a futuristic touch.

The present inclination to make information a problem looks for (new) technologies to fix it. However, new technologies do not necessarily fix the problem, because it stems from the ideological obsession both with quantity as precipitated by technology and, at the same time, with technology as the means of gaining control. My claim is that a critical history of information will strengthen current critiques that aim at debunking current ideological constructions of information.

95 For WHO's use, see https://www.who.int/docs/default-source/coronaviruse/situation-reports/20200202-sitrep-13-ncov-v3.pdf (2/4/2020). In distinguishing between information, misinformation, and disinformation, I consider information to be alethically neutral. For more about the distinctions between this trichotomy, see Søe, Urge to Detect.

Federal administration in the lead-up to the computer

Nick Schwery

Abstract

This paper examines the context in which the first computers, yet to be ordered in fall 1960, were interpreted by the Swiss federal administration. The focus of investigation is the moment when, by means of an inconspicuous note to the Swiss Federal Council, responsibility for upgrading the federal administration's Statistical Office to computers shifted from the Statistical Office to the Central Office for Organizational Affairs. This paper will further show how the Central Office was able to define the computer not as a better tool for statistics (in line with punched card machines), but rather as a general-purpose machine available to the entire federal administration. Henceforth, computer projects would inevitably involve a series of trade-offs: between general availability and individual use, between management hierarchies and project organization, between the needs of the computer and the needs of the entire federal administration, and between a range of actors who had to find ways to cooperate.

Introduction

This article aims to describe the initial interactions between computers and the Swiss federal administration.[1] In so doing, it will argue against the commonly held idea that technology enters public administration ready to be used – as a solution to a specific problem, for example, growth – and precipitates change. Were that the case, the administration would have to adapt to the new technology.[2] I will show, on the contrary, how negotiations regard-

1 I would like to thank the reviewers and editors for their valuable input. All quotations in this article have been translated by the author.

2 For example, Arre Zuurmond, From Bureaucracy to Infocracy: Administrative Reform by Technological Innovation in the Netherlands, Baden-Baden 1997; Rüdiger Bergien, «Big Data» als Vision. Computereinführung und Organisationswandel in BKA und Staatssicherheit (1967–1989), in: Zeithistorische Forschungen/Studies in Contemporary History, 2017, 14(2): 258–285; or Guido Koller, Sebastian Schüpbach, Geschichte der

ing the use and integration of computers in the Swiss federal administration took place before the first computer system was ever ordered. As Atshushi Akera has shown for the United States, the computer had to be interpreted as a tool for administration.[3] Rules, procedures, and regulations all had to be defined, a computerized administration imagined, and infrastructure organized even while computers were still in the offing. This early history of computerization shows how the federal administration negotiated early computer projects and how it planned to digitize processes and procedures between bureaucratic, programming, and project forms in the near future.[4]

In Switzerland, the prospect of a computerized administration began to emerge in October 1958, when 51-year-old Roger Bonvin, a member of both the Swiss National Council and the Christian Democratic People's Party, submitted a *kleine Anfrage*, or written question, to the Federal Council. Bonvin was an ambitious politician from the canton of Valais. His question expressed dissatisfaction. In his view, the Federal Statistical Office did not «meet the requirements of today's life» – neither organizationally nor with

modernen Verwaltung, 2016: http://www.livingbooksabouthistory.ch/de/book/the-history-of-modern-administration (8/12/2020): «The growth of the federal administration meant that more and more information had to be processed in ever shorter time. With this background, technological change began to have a lasting impact on the federal administration» (p. 5).

3 Atshushi Akera, Engineers or Managers? The Systems Analysis of Electronic Data Processing in the Federal Bureaucracy, in: Agatha C. Hughes, Thomas Parke Hughes (eds.), Systems, Experts, and Computers: The Systems Approach in Management and Engineering, World War II and After, Cambridge 2000, pp. 191–220. Akera points out that before the computer came into the administration, it first had to be interpreted as an administrative tool. In America, this was done by the National Bureau of Standards in the person of Samuel Alexander. Akera identifies areas of tension between technicians and management: «[A]ny analysis of data processing operations could become an implicit critique of the existing bureaucracy» (p. 202). See also Nick Schwery, Die Maschine regieren. Computer und eidgenössische Bundesverwaltung, 1958–1965, in: Preprints zur Kulturgeschichte der Technik, 2018 (29).

4 The issue is not only about how the computer came into the federal administration; it is also about how the federal administration in turn migrated into the computer and how computer projects were carried out. See David Gugerli, Wie die Welt in den Computer kam. Zur Entstehung digitaler Wirklichkeit, Frankfurt am Main 2018.

respect to the pace of work.[5] He took aim at four specific areas: first, the inadequate «internal structure» and cooperation with other departments within the federal administration; second, the lack of «rationalization of the administration's working methods»; third, the lack of «electronic mechanization of work processes»; and fourth, the non-existent adaptation to «the new needs».[6] Bonvin's critique was formulated in abstract terms, but his meaning was clear: the Federal Statistical Office needed to modernize.[7] Bonvin had trained in engineering at the Federal Institute of Technology Zurich (ETH Zurich), and had been involved in constructing the first Dixence dam (1932–34) and the Mauvoisin dam (1949–55). He entered politics at the age of 41 as a «conservative-Christian-Socialist local councilor of Sion». His hope was that modernizing the statistical office through electronic mechanization would also make it more efficient.[8] To do that, the structures of the office would have to change. Bonvin must have been quite determined on that point, because in his request, he followed up his critique by asking the Federal Council whether they «did not consider it necessary to draw up a reorganization plan».[9] For Bonvin, modernizing the administration was unthinkable without such a plan.

---

5 Swiss Federal Archives (BAR) E3320B#1976/141#1*, Kleine Anfrage Bonvin, 1/10/1958.

6 Ibid.

7 Eduard, head of the Institute of Applied Mathematics at ETH Zurich, was also involved in the Mauvoisin dam. At his institute he calculated the deformation under water pressure of the dam under construction. See Walter Gautschi, Schweizer Expats in den USA, 2016: https://ethz.ch/content/dam/ethz/special-interest/infk/department/Images%20and%20Content/Spotlights/Gautschi_Walter_Expats.pdf (16/7/2020). Stiefel, Switzerland's computer pioneer, put Konrad Zuse's Z4 calculating machine into operation at ETH in 1950 and, with the collaboration of Ambros Speiser and Heinz Rutishauser, built the Ermeth computer – «the first Swiss computer» – which he operated until 1963. See Hans Neukom, Early Use of Computers in Swiss Banks, in: Annals of the History of Computing, IEEE, 2004, 26(3): 50–59; Evelyn Boesch Trüeb, Eduard Stiefel, in: Historisches Lexikon der Schweiz (HLS), 2010, https://hls-dhs-dss.ch/de/articles/031672/2010-11-29/ (16/7/2020).

8 Georges Andrey, Roger Bonvin, in: Historisches Lexikon der Schweiz (HLS), 2009, https://hls-dhs-dss.ch/de/articles/004723/2009-12-10/ (29/4/2020).

9 BAR, Kleine Anfrage Bonvin.

Four short years later, the Statistical Office put into operation the Swiss federal administration's first computer system. It was based in a computer center (*Rechenzentrum*) set up especially for the new computers under the umbrella of the Statistical Office, which had to cooperate with other departments. Achieving the technological «state of the art» fulfilled the promise of modernization. Bonvin's four criticisms had been addressed, and Bonvin himself had a seat in the Federal Council. But it wasn't only the Statistical Office that had changed.

Renewing the equipment

In the meantime, the Statistical Office had already begun to think about renewing their equipment for the 1960 census. Bonvin's *kleine Anfrage* aside, the administration, and with it the Statistical Office, had a perpetual improvement problem: under constantly changing circumstances, «the willingness and ability to reform and to make other adjustments is still among the core characteristics of a good administration».[10] If the administration wanted to be a good administration and the Statistical Office a good office within this administration, it constantly had to adapt. But reform was never straightforward, and the office was forced to fall back on allies. As a result, proposals and reports regarding «renewal of equipment» typically included Bonvin's request as the starting point, which lent a note of legitimacy to the Statistical Office's efforts.[11] Equipment renewals had been subject to review at regular intervals of ten years for each new census since the 1920s.[12] But for

10 Wolfgang Seibel, Verwaltung verstehen. Eine theoriegeschichtliche Einführung, Berlin 2017, p. 102.

11 For example, in the final report with a proposal for upgrading the equipment: «On 1 October 1958, Bonvin submitted a ‹*kleine Anfrage*› concerning the reorganization of the Statistical Office, in which he asked, among other things, whether the ‹efficiency of the office might not be considerably increased by electronic mechanization of the work processes for faster evaluation of all material›». BAR E6502-02#2002/226#16*, Report on Renewal of the Equipment ESTA, 8.1960.

12 From 1850 to 2000, the census provided information on the population, households, buildings and dwellings in Switzerland every ten years. Since 2010, the Federal Statistical Office has conducted the census annually. In order to make the process easier for people,

the 1960 census, and for the first time, computers were being considered in addition to the punched card technology used for decades.[13] This is hardly surprising, given the radically widening scope for the use of computers at the end of the 1950s. Computers were no longer restricted to the military and to science but were also being used in business and administration.[14] As early as 1951, a UNIVAC computer from Remington Rand was in use by the US Census Bureau, and in 1957 the first computers found their way into German public administration to process the annual payroll tax adjustment.[15] Computer manufacturers boasted that their new machines could further rationalize mass data processing, which was often already being done with punched card systems. These mostly classical administrative tasks typically involved processing statistical data, preparing pay slips or stock inventories, executing accounting tasks and interest-rate calculations, and updating savings accounts.[16]

much of the information is taken from the population registers of the communes and cantons, the federal registers of persons, and the federal register of buildings and dwellings. See Bundesamt für Statistik, Volkszählung, https://www.bfs.admin.ch/bfs/de/home/grundlagen/volkszaehlung.html (15/12/2020).

13 The name given to the new technology, which was negotiated in the federal administration, was anything but certain around 1960: from electronic data processing machine (EDPM) to EDP system to computer, electronic computer, digital or electronic computer system, or simply calculator. In the article, I stick consistently, somewhat ahistorically, to the term *computer*, which came to be used in distinction to the so-called conventional punched card machines, which computers slowly began to replace.

14 See computer history, e.g., Martin Campbell-Kelly et al., Computer: A History of the Information Machine, Boulder 2014; Gugerli, Wie die Welt; Michael Sean Mahoney, Histories of Computing, Cambridge MA 2011.

15 See Hans Peter Bull, Verwaltung durch Maschinen. Rechtsprobleme der Technisierung der Verwaltung, Köln 1964, p. 37; Robert Garner, Early Popular Computers, 1950–1970, 2015, https://ethw.org/Early_Popular_Computers,_1950_-_1970#Citation (29/4/2020) or Ricky Wichum, Verwaltungsrecht und Automation um 1960, in: Dennis-Kenji Kipker et al. (eds.): Der normative Druck des Faktischen: Technologische Herausforderungen des Rechts und seine Fundierung in sozialer Praxis, Stuttgart 2019, pp. 69–87.

16 See Josef Egger, «Ein Wunderwerk der Technik». Frühe Computernutzung in der Schweiz (1960–1980), Zurich 2014.

Bonvin's *kleine Anfrage* took advantage of this widening of the computer's scope of application. The Swiss federal administration, too, should benefit from the electronic mechanization of work processes. In the Statistical Office, the intended «renewal» was approached in the usual bureaucratic manner. Because the equipment review took place every ten years, the office was able to draw on a certain amount of experience and routine. IBM and Bull, both computer manufacturers that had business relations with the federal administration through punched card systems, offered their new IBM 7070 and Bull Gamma 60 systems. The Statistical Office examined the offers and ran through the census tasks with both manufacturers.[17] The office then attempted to calculate the economic efficiency of these computers compared with «conventional» punched card systems, assuming a lifetime of twelve years. According to the calculation, the massively higher purchase price of the computer should be compensated by savings on personnel. A solution using the cheaper of the two computers was expected to cost around 7 million Swiss francs[18] – about 3 million francs less than calculations based on punched card systems.

The Statistical Office decided on the IBM system, «not only because of the price, but also because it is a company whose conventional machines we have been using for 30 years, so we have a well-established relationship with them».[19] Bull also operated punched card systems in the federal administration, but not in the Statistical Office. The decision favoring IBM was more cost-effective and also promised continuity. The Statistical Office remained an IBM power user. Perhaps it was due to this well-established cooperation that little attention was paid to the transition from punched card systems to

17 «In 1958, International Business Machines Corp. (IBM), supplier of punched card machines to the Statistical Office, introduced to the market a sensationally innovative medium-sized EDP device known as the IBM 7070. Soon afterwards, BULL offered a similar device, the Gamma 60.» BAR E6502-02#2002/226#16*, Report on Renewal of Equipment, Statistical Office, 31/8/1960. An offer from computer manufacturer Remington Rand reached the federal administration too late.

18 How approximate this calculation was is obvious from the different stages of the report, which can be found in the archive. The figures in the table had to be corrected several times.

19 BAR, Report on Renewal.

computers. It was apparently assumed that the computers would be seamlessly integrated into the office's existing structures. The introduction of the new computer technology was calculated at an additional cost of only 90,000 Swiss francs. Thus, on 31 August 1960, the Statistical Office put in a request to the Federal Council that «the electronic data processing system IBM 7070 [coupled with an IBM 1401] be ordered immediately» and added: «Because of the delivery periods of 1–1 1/2 years, a decision should be made quickly».[20] By the end of 1961 at the latest, the computer system should be ready to process the punched cards resulting from the census and output the data as statistics.[21]

Negotiating computer use

But the Statistical Office had not expected the Department of Finance to intervene. Despite the continuity ensured by the decision to go with IBM and despite agreement that automation was needed in the federal administration to save money and personnel, the Statistical Office's request hit a roadblock. A handwritten note accompanying the report stated concisely: «For a transaction of such a costly scope, only a decision by the Federal Council is possible»[22] Federal Councilor Hans-Peter Tschudi, a Social Democrat and head of the Federal Department of Home Affairs (*Departement des Innern*), added a second note addressed to the Statistical Office, which was under his authori-

20 BAR, Report on Renewal.

21 The census was conducted in the following way: In 1960, census takers armed with forms went door to door throughout Switzerland. The completed forms were sorted at the Statistical Office, and the information was transferred to punched cards. The cards were then fed into the computer, which processed the information into statistics. According to a Swiss film newsreel from 1961: «Tons of completed census forms are now arriving at the Federal Statistical Office in Bern. Here is Switzerland, distilled on paper. […] The sorting and extracting of information have only just begun. But in a year's time, the counters will have worked their way through the mountain of paper to prepare the forms for processing by huge electronic machines.» Schweizer Filmwochenschau, Ausgefüllte VZ-Formulare kommen ins BFS zurück, 1961, https://www.youtube.com/watch?v=eHwumxwUHT8 (9/7/2020).

22 BAR E6502-02#2002/226#16#*, Note, 2/9/1960.

ty: «Request for a motion to the Federal Council. Considering its importance, an expert opinion from the Central Office for Organizational Affairs should be attached.»[23] The requested upgrading of the equipment with computers required an expert opinion from the federal administration's Central Office for Organizational Affairs (*Zentralstelle für Organisationsfragen der Bundesverwaltung*). Tschudi was looking for support for the high investment costs and wanted to politically secure the plans of one of his offices. With this note to the Federal Council, the computer requested by the Statistical Office was directly linked to the organization of the federal administration. The future use of computers implied more than the rationalization, cost savings, and modernization of one office. From this point on, the computers in the administration were tied to the purpose of administrative reform.

The federal administration's Central Office for Organizational Affairs had been set up by the Federal Council in 1953 to slow down the politically controversial expansion of the welfare state after the Second World War.[24] The creation of the Central Office was initially an internal response to failed reform attempts using staff surveys and external experts after the war; a year later, the office would be a response to the «popular initiative for federal administrative control» (*Volksbegehren betreffend einer eidgenössischen Verwaltungskontrolle*).[25] Switzerland was experiencing strong population growth

---

23 Ibid.

24 In Switzerland, the 1950s represented one of two decades of growth in the postwar period. In the long 1950s, «economic growth, social prosperity and individual welfare [still] went hand in hand», as economic historian Jakob Tanner pointed out in 1994 in a comprehensive review (p. 39). According to Tanner, ambivalence toward growth, for example, the disappointment and frustration structurally inherent in economic expansion – the «paradox of plenty» – and the lack of or poorly trained or unskilled human capital only entered the «collective horizon of perception» during the 1960s (p. 38 f.). The internal view of the federal administration in Switzerland complements Tanner's overview and shows that this ambivalence began to show itself earlier in the federal administration than in production. See Jakob Tanner, Die Schweiz in den 1950er Jahren. Prozesse, Brüche, Widersprüche, Ungleichzeitigkeiten, in: Jean-Daniel Blanc, Christine Luchsinger (eds.), Achtung: die 50er Jahre! Annäherungen an eine widersprüchliche Zeit, Zurich 2004.

25 See Koller, Schüpbach, Geschichte moderner Verwaltung and BAR E6500-02#1986/114#193*, Evaluation Sparexpertisen, without author or date. In response to the popular initiative, the department changed its name from «Coordination Office for Savings and

and increasing purchasing power, and the Swiss were committed to the «American way of life».[26] The problem of growth within the administration was to be coordinated by the Central Office. Its mandate was to «continuously examine the appropriateness and effectiveness of the organization and working methods of the federal administration as well as the possibility of making it more economical».[27] Administratively, the Central Office was subordinate to the Department of Finance and Customs. Technically, however, it was largely independent and directly responsible to the entire Federal Council, which also appointed its head.[28] Its first director was Otto Hongler (1907–1988), who earned his doctorate in economics and taught at the ETH Institute of Management (*Betriebswissenschaftliches Institut*). Before moving into public administration, Hongler had worked as an expert on organizational issues in trade and industry.[29] The Central Office for Organizational Affairs was a small office, in line with its mission. From an initial two civil servants, the Central Office grew to five by 1960. In its first few years of operation, the office mainly produced or commissioned expert reports. It also tried to obtain input to improve the organization of the federal administration in a decentralized manner with the help of *Organisationsmitarbeiter* (organizational assistants), who were appointed in the departments as links to

Rationalization Issues» (*Koordinationsstelle für Spar- und Rationalisierungsfragen*) to Central Office for Organizational Affairs of the Federal Administration. See BAR E6502–01#1993/126#246*, Report on Present-Day Efforts to Organize Work Expediently and Economically, Hongler, 1956.

26 Jakob Tanner, Zwischen «American Way of Life» und «Geistiger Landesverteidigung». Gesellschaftliche Widersprüche in der Schweiz der fünfziger Jahre, in: Unsere Kunstdenkmäler, 1992, 43(3): 351–363.

27 BAR E6502–01#1993/126#254*, Federal Act on the Central Office for Organizational Issues of the Federal Administration, 6/10/1954.

28 See Peter Olivet, Die Organisation der Organisation der öffentlichen Verwaltung in der Bundesrepublik Deutschland. Aufbau und Arbeitsweise der zentralen Organisationsstellen in der öffentlichen Verwaltung der Bundesrepublik Deutschland, Berlin 1978, p. 254 ff.

29 See Andrea Weibel, Otto Hongler, in: Historisches Lexikon der Schweiz (HLS), 2005, https://hls-dhs-dss.ch/de/articles/011524/2005-02-08/ (1/5/2020). Hongler remained director of the Central Office for Organizational Affairs until his retirement in 1973.

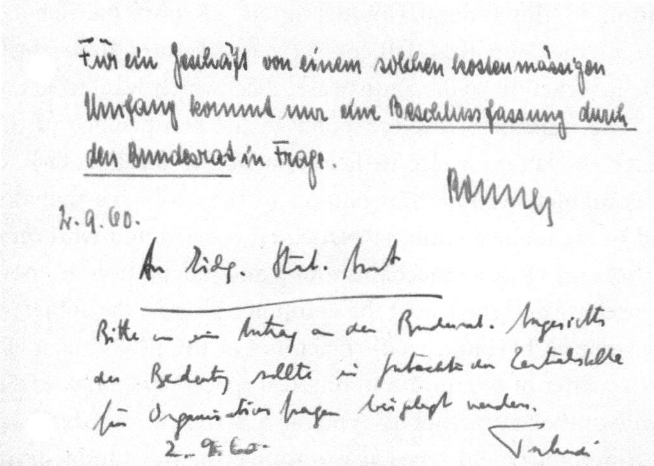

Für ein Geschäft von einem solchen kostenmässigen Umfang kommt nur eine Beschlussfassung durch den Bundesrat in Frage.

2.9.60.

An Eidg. Stat. Amt

Bitte um einen Antrag an den Bundesrat. Angesichts der Bedeutung sollte ein Gutachten der Zentralstelle für Organisationsfragen beigelegt werden.

2.9.60.

Tschudi

Fig. 1: Handwritten notes in the «Report on Renewal of the Equipment of the Statistical Office» (Swiss Federal Archives E6502-02#2002/226#16*). «A transaction of such costly dimensions requires the approval of the Federal Council. 2/9/60 [illegible signature] / To Federal Statistical Office / Request for a motion to the Federal Council. Considering its importance, an expert opinion from the Central Office for Organizational Affairs should be attached. 2/9/60. Tschudi» BAR, Note.

the Central Office. The organizational assistants reported progress and provided suggestions regarding rationalization of their respective departments to the Central Office on an annual basis. The review of the organization of the federal administration, its departments and offices, and its working methods was thus partly delegated to the departments and their offices and divisions. The Central Office saw itself as an economically savvy advisory center for rationalization and it nominated civil servants, rather than an administrative elite to initiate the changes. In nominating and working with the organizational assistants to initiate change, the Central Office was targeting as their change agents not an administrative elite, but civil servants.

According to Tschudi's note of fall 1960, whether and how the Statistical Office's equipment was to be renewed was suddenly no longer a matter for the Statistical Office. Rather, the matter was the responsibility of the Central Office for Organizational Affairs, which was charged with making a recommendation to the Federal Council in its «Report on Renewal of the Equipment of the Statistical Office».[30] The computer to be ordered had escaped the jurisdiction of the Statistical Office, which wanted to control and operate it. The transfer of the computer to the competence of the Central Office offered the latter a welcome field of action to fulfill its task. Only then could the computer become the solution to the problems that Bonvin had formulated in his request (internal structure, cooperation with other offices, rationalization, electronic mechanization, and adaptation to new needs). Both the purchase and the use of the computer became the subject of negotiations that went far beyond specific machines or the 1960 census. Thereafter, it was also a matter of organization, physical space, personnel, and coordination not only of the Statistical Office but also of the entire federal administration. The computer would serve as a resource for the administration to reconfigure itself under the guise of rationalization. There was no need even for a reorganization plan for the Statistical Office, as Bonvin had thought in 1958.[31]

Computer? Yes, but ...

The requested expert report triggered action within the Federal Administration. Otto Hongler personally took over responsibility for it. He quickly sent a catalogue of questions comprising eleven points to the Statistical Office, which had to provide answers and documents.[32] In addition, he consulted various departments within the federal administration and punched card and

30 BAR, Report on Renewal.

31 This was apparently the case wherever public administration and computers came into contact. In 1966, Niklas Luhmann wrote on the subject of the computer in public administration: «A refreshing thought-provoking impulse comes from the fortunate fact that the machines are so expensive. Their price forces one to rationalize the organization of data processing outside the actual system to an extent that would have been unfeasible without this impulse.» Niklas Luhmann, Recht und Automation in der öffentlichen Verwaltung. Eine verwaltungswissenschaftliche Untersuchung, Berlin 1966, p. 9.

32 BAR E6502-02#2002/226#16*, Request of the Central Office to the Statistical Office Renewal of Equipment, 15/9/1960.

computer companies. As a result of the investigation, he announced that «the acquisition of a powerful electronic data processing system (EDPM) [was] appropriate for the federal administration» – but only if adjustments were made.[33] The Central Office presented the computer to be ordered not as a tool for the 1960 census but as a flexibly applicable piece of administrative equipment – a general-purpose machine.[34] This reinterpretation necessitated adjustments to the configuration of the system, personnel, and access.[35]

The first adjustment to be made – configuring the computer system – was occasioned by the different demands placed on the computer. While the Statistical Office wanted a computer that could be used to calculate the 1960 census more quickly, the Central Office wanted an administrative machine that could be used more flexibly and whose field of application extended beyond statistics. For this reason, Hongler recommended supplementing the IBM system with additional components that were «necessary for carrying out work for other departments».[36] These additional components meant almost 800,000 Swiss francs in additional costs, bringing the purchase price of the system to 5.2 million Swiss francs. In addition, Hongler calculated other machine costs during the period of operation amounting to 2.5 million Swiss francs. In parallel, suitable premises had to be found for the computers, which required 240 square meters of space.[37] After a renewed economic efficiency calculation, which now considered «the total costs of the punched card or EDP service», the savings compared with the punched card operation amounted to 2.8 million Swiss francs.[38] The costs were thus minimally lower than in the calculation made by the Statistical Office. Despite higher machine costs, the economic efficiency remained assured.

33 BAR E6502-02#2002/226#16*, Expert Report Hongler, 2/11/1960.

34 Only at this point was structural equality between computer and administration a possibility, as described by John Agar, The Government Machine: A Revolutionary History of the Computer, Cambridge MA 2003.

35 BAR, Expert Report Hongler.

36 BAR, Expert Report Hongler.

37 The machine room of the Statistical Office was not suitable for this purpose because it was not high enough. See BAR E6502-02#2002/226#16*, IBM7070/1401, 26/9/1960 and BAR, Expert Report Hongler.

38 BAR, Expert Report Hongler.

Hongler identified a second critical adjustment in the transition of work processes from punched card systems to computers. The ongoing work at the Statistical Office had be transferred to the computer as quickly as possible, and not, as originally intended in the Statistical Office report, after the end of the census in 1965. Because the census would require «employment of a larger number of programmers» in the first three years, «the use of an EDP system [...] is largely a personnel problem.»[39] These yet unidentified personnel – the roles of the programmer and analyst were anything but clearly defined in 1960 – were supposed to transform the general-purpose computer into an administrative machine and specialize it. In his expert report Hongler therefore requested «that the selection of personnel be given the greatest attention». Without «competent people with a talent for these tasks», the computer could be used neither «properly» nor in a «cost-saving» manner.[40] The personnel problem posed by the question of programming and integrating technical staff into the administrative structures would accompany the computer in the administration for the next decades.[41]

The third adjustment appeared to have been more sensitive, to the extent that Hongler wanted to come back to it «in a later, separate report». The issue of «coordination between departments, responsibility for controlling the use of punched cards and EDPM», i. e., the question of access to computers, could possibly risk the consensus of the first report that the «computers were adequate».[42]

39 Ibid.

40 Ibid.

41 Programming the computer was a common problem for all employers at the time and was accompanied by the problem of general staff shortages and skills deficits. A closer look at the figure of the programmer exemplifies these issues. See David Gugerli, Der Programmierer, in: Alban Frei, Hannes Mangold (eds.), Das Personal der Postmoderne, Bielefeld 2015, pp. 17–32; Nathan Ensmenger, The Computer Boys Take Over: Computers, Programmers, and the Politics of Technical Expertise, Cambridge 2010, or JoAnne Yates, Structuring the Information Age: Life insurance and Technology in the Twentieth Century, Baltimore 2008.

42 BAR, Expert Report Hongler.

Centralized coordination

Up to this point, the Central Office had little trouble making its case. A few adjustments to the configuration here, some hints at possible difficulties there. It was clear that the administration wanted computers. In the separate «Report and Motion for the Creation of a Unit Concerning the Coordination of Punched Card and EDPM Use in the Federal Administration», the chain of argumentation was extended.[43] According to rumors, the Central Office knew to report that the acquisition of computers was an issue not only in the Statistical Office but also in the Alcohol Board (*Alkoholverwaltung*), the Treasury and Accountancy Service (*Kassen- und Rechnungswesen*), the Printing and Supplies Office (*Drucksachen- und Materialzentrale*), and in various companies of the War Technology Division (*Kriegstechnische Abteilung*).[44]

The Central Office for Organizational Affairs therefore assumed that «a breadth and depth development [was] emerging throughout the administration in the punched card and EDP area which [could not] be left to itself.»[45] The computer system that the Statistical Office wanted to acquire in renewing its equipment was made the central focus of an «inescapable trend» by the Central Office.[46] The administration could neither stop the computers nor leave them to themselves. The reason for this was that in comparison to punched card machines, the computers «would enable and categorically demand a more intensive cooperation between the punched card and EDP services».[47] Whereas in the first report Hongler still stressed the usefulness of computers for the federal administration, by the second report the decision in favor of the computer had already become an inevitability.

The mantra of the Central Office was clear: every department, every position in the administration should be able to access computers, whether in the Alcohol Board or the War Technology Division. But due to their high

43 BAR E6500-02#1986/114#74*, Report and Motion on the Coordination of Punched Card and EDPM Use, 28/11/1960.

44 Ibid.

45 Ibid.

46 Ibid.

47 Ibid.

price, the computers had to be used at a central location in a computer center. The Central Office wanted to prevent each department from evaluating, ordering, and operating its own machines. The Central Office framed the computers to perform balancing acts between general availability and individual use. In other words, the resources of the general-purpose machines had to be made available to the individual needs of the departments. The problem that the computer represented in this interpretation was the problem of access. How were departments to access computing power in the computer center?[48] The solution lay in bureaucratically regulating the coordination of the administration's computers.

For this purpose, Hongler first analyzed the actual situation, then critiqued it and transformed the critique into a «proposed solution for the reorganization of the punched card services of the federal central administration». How would the Central Office do that? It disavowed the «isolated action of the departments», as the Statistical Office had done until the intervention of the Department of Finance. In the future, such an approach would «no longer [be] responsible». «The tasks, competencies and responsibilities in the area of punched card and EDP deployment of the federal administration» had to be redefined: «It is therefore necessary to propose ways and means that will allow the sum of all activities of the punched card offices to be optimally designed.» For the Central Office it was clear that the computers had to be operated centrally in a computer center – under the umbrella of the Statistical Office but as a service provider for the entire administration. According to the Central Office, to make the computers usable for other de-

48 The problem of access to computing power was widespread around 1960. At the Massachusetts Institute of Technology (MIT), the problem of accessing the large central computer led to the development of time-sharing. Various users could access computing power via decentralized terminals without having to stand in line at the computer. Key to this development was the supervisor, a program that controlled all activities and could interrupt one procedure in favor of another, so that all users had access to expected computing power. The problem of access was technically solved at MIT, as the supervisor distributed the central computing capacity fairly to the decentralized users. See Fernando J. Corbato et al., An Experimental Time-Sharing System, in: AIEE-IRE '62 (Spring) Proceedings of the May 1–3 Spring Joint Computer Conference, 1962, 335–344.

partments, coordination «had to be created, planned, and ensured to a large extent».[49]

Hongler ruled out self-coordination among the departments in order to achieve the optimal design of the organization from a functional and economic perspective. The «natural tendency» for the departments to try to maintain «their autonomy and independence», he argued, was too great. He was concerned that only «the circumstances of one's own department» would be considered.[50] This would make the computer too individualized, and only a few would be able to access it. For the same reasons, Hongler spoke out against the formation of a commission. The interests of the departments would not be compatible with the task of coordination. Neither self-coordination nor a commission was a workable solution for organizing the operation of the computers in the federal administration. What would be, then? Hongler proposed an office which, first, would have to have «an overview of the entire punched card services of the federal administration»; second, would have to know «the problems and plans of the individual departments»; third, would have to follow «the progress of automation technology»; and, finally, would have to plan and coordinate «the development of punched card services in the interest of the entire administration over the long term».[51]

The modernization of the administration should be accompanied by technocratic centralization. The report does not specify where the office should be set up, but states that it would be a demanding task: «This office must have thorough knowledge of organizational matters and of punched card and EDP technology, and must be equipped with the necessary skills to perform its task. Its coordinating role also requires psychological skills and experience in proposing and implementing organizational changes».[52] The requirements focused on organizational affairs and, more specifically, on «proposing and implementing organizational changes» and knowledge of computer technology. In addition to Hongler, Hans Kurt Oppliger, a 33-

49 BAR, Report and Motion.

50 Ibid.

51 Ibid.

52 Ibid.

year-old former employee of Bull, had also been working in the Central Office since 1957.[53] It must have been obvious that Hongler had defined a new office for which only the Central Office could be responsible.

On 8 December 1960, the reports and motions were summarized. The computers for the census could no longer be negotiated without the creation of a «coordination office for automation».[54] On 16 December, the Federal Council dealt with the case and decided to order the IBM 7070 and 1401 machines.[55] Although the question of location was still unresolved – the Statistical Office, the Construction Directorate and the Central Office were still to negotiate the space – the computer center was «provisionally» placed under the control of the Statistical Office. The final subordination should be clarified by the Central Office «with the interested parties», and a report and application should again be submitted to the Federal Council.[56] In addition, the Federal Council decided that «the planning, monitoring, and promotion of the activities of all punched card and EDP services of the Federal Administration [...] should be transferred to the Central Office for Organizational Affairs as a coordinating office».[57] Oppliger became its first head. The way of operating around the by then only ordered computers was characterized by centralized coordination.

In this short period of time, the Central Office took advantage of the opportunity to exploit the tension between general availability and individual use of the computer in order to gain legitimacy. The Central Office was able to interpret the computer as a vehicle for gaining control and ultimately to better fulfill its task of reforming the federal administration. It had managed

53 See Sarah Brian Scherer, Hans Kurt Oppliger, in: Historisches Lexikon der Schweiz (HLS), 2008, https://hls-dhs-dss.ch/de/articles/011720/2008-10-23/ (9/4/2020).

54 BAR E6500-02#1986/114#75*, Report and Motion Summarized, 8/12/1960.

55 «The Print and Material Headquarters is authorized to order the EDP system IBM 7070-1401 and the necessary auxiliary machines – immediately, subject to the approval of the budget by the Federal Assembly». BAR E6502-02#2002/226#16*, Resolution Federal Council, 16/12/1960.

56 «The Central Office for Organizational Affairs of the federal administration clarifies the question of definitive subordination with the interested parties and then submits a report and proposal to the Federal Council». Ibid.

57 Ibid.

to negotiate and expand its own responsibility vis-à-vis the computer. The balancing act between the necessary specialization of the computer, for example, for the census in the Statistical Office, and the requirement that the expensive computer system be used to rationalize the entire administration offered the opportunity for this. With the expected «breadth and depth development» of computer use, control over computerization promised influence for the Central Office in the same dimension – across the breadth of departments to the depth of divisions and offices.[58]

Between project and line

The Central Office managed to establish the organizational framework for the implementation of the computer within the Swiss federal administration two years before the first computers were put into operation. By successfully linking the problem of modernizing to the computer, it was empowered to restructure the administration in the near future. The computers were not to be installed as part of the Statistical Office's equipment but rather in the federal administration's newly created electronic computer center – right next to the federal parliament building, in the courtyard of the Bernerhof, which had been rebuilt for this purpose.[59] Although the computer center remained organizationally under the umbrella of the Statistical Office, it was intended to provide service for the entire administration. From now on, any future use of computers had to be planned, monitored, and supported centrally. For this purpose, the specifications of the Central Office were extended. Its area of responsibility now also included the coordination of all the federal administration's efforts in the field of automation. For computer problems, and specifically for the problem of access, no other routines or protocols existed. These problems were the first of their kind. And they were solved, at least temporarily, on an organizational level between the Statistical Office, the Department of Finance, the Department of Home Affairs, the Federal Council, and the federal administration's Central Office for Organizational Affairs.

58 BAR, Report and Motion.

59 See Schwery, Die Maschine regieren.

In less than four months, between September and December 1960, the negotiation ground for the future use of computers was defined. Not in response to the computer, but in anticipating computer use and computer access from a holistic standpoint. Not as an answer to the growth of the administration, but in the context of it. The Central Office succeeded in defining the computer as a general-purpose machine. In doing so, it not only succeeded in finding a new strategy for future administration reorganizations but also significantly increased its influence. It virtually became the supervisor of the operating system, deciding what could be done with the computer, when, where, and by whom – without operating computers itself. The computer became the second, far more effective link between the Central Office and the departments, alongside the organizational assistants. The interaction between the administration and the computer was structured long before the first computers went into operation in spring 1962.

That same year, Roger Bonvin was elected to the Swiss Federal Council, making a name for himself as someone who cared about the Swiss cantons and regions.[60] Computers, however, disappeared from his agenda. In his *kleine Anfrage*, he had imagined computers to be used in already existing structures, as tools for change inside the Statistical Office. The computer was a vehicle for becoming and appearing modern. Bonvin thus stood for the «compromise formula» that had shaped Switzerland in the postwar period, in which the liberal ideology of (technological) progress combined with an ideology of cultural and structural preservation.[61] For Bonvin and his career, it did not matter that the first computers in the federal administration came into operation other than the way he had intended.

Like Bonvin, the Central Office also had to find a balance between change and preservation. Upcoming problems having to do with computers, for example, how to translate processing pay slips into a program to process them with computers, had to be reconciled with well-established line hierar-

60 He was well known for always standing up for the mountain cantons, for example, during the «Gotthardkreuz» railway project. In what became known as the «Furka affair», only the Furka base tunnel was ever completed, and even that only with technical difficulties and enormous budget overruns. See Andrey, Bonvin.

61 Mario König et al., Dynamisierung und Umbau. Die Schweiz in den 60er und 70er Jahren, Zurich 1998, p. 12.

chies. Programming computers was a project-based activity. The classic top-down chain of command was unavoidable, but it promised little success in these temporary spaces between the computer and the administration. In order to move administrative processes to the computer, programmers and analysts in the computer centers had to cooperate with departmental civil servants to bring knowledge of computer programming and of administrative workflows together in project groups between the lines. In this constellation, transferring processes according to a different logic – that of the computer and that of the administration – offered an option to change how things were done with the excuse that either the needs of the computer or the needs of the office automating a workflow required it. The computer became a vehicle for self-reflection, and subsequently a way to change the administration and the way it did its work under the radar of politics. Because cooperation between highly heterogeneous institutions and actors stood at odds with the line organization, the Central Office and its directives had to find ways to regulate the interactions in a way that continued to respect the decision-making of the line hierarchies. Within the context of the computer, certain new project structures and a new type of the old, bureaucratic civil servant were recognizable. The buzzwords were «general-purpose» and «individual use», «programming» and «analyzing», «center» and «coordination». In the early interactions between the computer and the administration, a testing ground opened up around these buzzwords for heterogeneous actors to find ways to cooperate and to manage projects. A nascent digital federalism had to link up with the burgeoning routines and standards inside this organizational framework. These were routines and standards for managing, initializing, and unwinding computer projects in different phases, at three different levels of cooperation between heterogenous actors – the totality of computer projects within the federal administration, a computer project as a unit (including the definition of objectives), and execution of an already defined computer project.

Within the federal administration, the next big question after ordering the first computer in 1960 became how to transform the general-purpose machine into a specialist for doing work for the federal administration. The problem of programming and programmers, at the intersection of computer and organizational structure, loomed on the horizon.

The promise of an automated migration policy: on planning an information system in the Swiss federal administration in the 1960s *

Moritz Mähr

Abstract

This article examines the planning of the Central Aliens Register in Switzerland from 1964 to 1971. This first national information system was designed to support the migration policy of the federal administration in such a way that foreign workers were distributed fairly among the cantons according to economic and demographic needs. Although it only fulfilled its original purpose to a limited extent, within a few years it became an important tool of the Federal Aliens Police and remained in operation until 2008. From the outset, this new sociotechnical infrastructure led to a conflict between the federal status quo and an automated, centrally controlled migration regime. The article analyzes how the federal migration regime in Switzerland responded to the tension between automation and federalism.

> Since the commission began its work, the nature and scope of immigration have changed considerably. The problem of foreign workers has also undergone a fundamental change in public opinion. While previously the large influx of foreign workers was widely regarded as desirable, even necessary, because it was conducive to the growth of our economy, recently there has also been a growing awareness of the disadvantages and dangers of this development.[1]

Introduction

In a period of strong economic growth in the 1950s and 1960s, Switzerland faced a complex mix of political developments. The economy promoted a massive influx of labor migrants, and cantons issued work and residence per-

* Many thanks to Lucas Federer, Henrike Hoffmann, and Philipp Krauer for their valuable comments on the manuscript.

1 Bericht der Studienkommission für das Problem der ausländischen Arbeitskräfte, Bern 1964, p. 8 (author's translation).

mits accordingly. Between 1960 and 1965 alone, the share of foreigners among the population rose from 10% to almost 15%. The steep rise fueled fears of wage dumping among trade unions and of «over-foreignization» (*Überfremdung*) among right-wing nationalists. Meanwhile, Italy, the country of origin of most migrants, increased pressure on Switzerland to improve poor working conditions and access to the social security system. Right-wing nationalists attempted to radically cap the percentage of foreigners in the population through a series of popular initiatives. The chances of success of these initiatives were good, and many employers feared the economic consequences. In order not to jeopardize the supply of skilled labor, the Federal Council increasingly intervened in the competencies of the cantons and attempted to centrally control the migration regime.[2]

The migration regime of the 1960s was a federal patchwork. Under the provisions of the Swiss constitution, cantons could issue work, settlement, and residence permits at their own discretion. Cantons with seasonally fluctuating labor demands (such as tourism, the construction industry, and agriculture) issued short-term permits valid for only a few months. At the end of the season, these workers would go back to their families and home countries before returning to Switzerland for the following season. Living and working conditions were often precarious: there was no statutory unemployment insurance, no minimum wage, and no support for integration. Cantons with a high demand for skilled workers issued longer-term work and settlement permits, allowed family reunification, and ensured swift integration.

The migration authorities were also very heterogeneously structured. In the canton of Basel City, where demand for skilled workers was high, the various authorities were combined in a well-organized office; in other cantons, authorities were involved at the cantonal and municipal levels. These differences meant that it was difficult to collect consistent statistics on the labor market and demographics at the national level. Some cantons reported figures to the administration from the cantonal alien police offices or employment offices, while others reported figures from resident registration of-

2 Peter Hablützel, Peter Gilg, Beschleunigter Wandel und neue Krisen (seit 1945), in: Beatrix Mesmer et al. (eds.), Geschichte der Schweiz und der Schweizer, Basel 1986, pp. 821–891.

fices. There was no clear overview that would allow coordination of the cantonal migration regimes. With the new nationwide restrictive measures introduced in 1970, the Federal Council broke with the status quo and upset the balance between the federal administration and the cantons. An automated information system, in which all aliens were to be recorded, was to statistically underpin the new policy and give it more legitimacy.

This article examines the planning of the Central Aliens Register (ZAR) information system from 1964 to 1971 to show how federal migration policy in Switzerland dealt with the tension between the federal status quo and automation. Sources from the Federal Archives in Bern are examined. They document the negotiation process between the federal administration and the cantons in this information infrastructure project. Our research shows that the historical development of Swiss migration policy and related government actions requires analysis of technical and organizational artifacts. Two concepts are of central importance: migration regimes and infrastructure. According to Lucassen, Hoerder, and Lucassen, migration regimes are a set of institutions and policies designed to restrict or facilitate the spatial movement of populations. Star defines information infrastructures as sociotechnical systems that record information about government actions and decisions and make it accessible to selected parties.[3]

Gees has shown that labor migration was a driver of European integration in the 1960s and 1970s. Switzerland was integrated into the Western European migration system through international treaties and active participation in international organizations. In his study of the restrictive migration policy measures, Mülli showed that the negotiation processes of Switzerland's federal migration regime resulted in new, highly technical governancet techniques. Espahangizi and Mähr have shown that these new techniques were largely based on statistical and computerized procedures.[4]

3 Jan Lucassen et al., Terminologien und Konzepte in der Migrationsforschung», in: Klaus J. Bade et al. (eds.), Enzyklopädie Migration in Europa: vom 17. Jahrhundert bis zur Gegenwart, Paderborn 2007, p. 39; Susan Leigh Star, The Ethnography of Infrastructure, in: American Behavioral Scientist, 1999, 43: 387.

4 Thomas Gees, Die Schweiz im Europäisierungsprozess. Wirtschafts- und gesellschaftspolitische Konzepte am Beispiel der Arbeitsmigrations-, Agrar- und Wissenschaftspolitik, 1947–1974, Zurich 2006; Michael Mülli, Kontingentierung von Migration.

This paper ties in with the history of computing and history of administrations in nation-states. As Agar has shown for Great Britain and Fleischhack for the Federal Republic of Germany, information systems played a key role in the administrative activities of the 1960s. For Switzerland, little research has been done on the interaction between computers and administration. Schwery has shown how the Federal Statistical Office acquired the first computer for the census in 1960 as an unspectacular replacement investment for conventional punched card machines, and how the Federal Computing Center developed from this acquisition. Zetti uses the example of the introduction of electronic data processing in the PTT (postal services, telegraphy, and telephony) companies at the end of the 1960s to show how staff associations, project management, the Computing Center, and the general directorate all worked together on the introduction of the computer to promote their own agendas. Gugerli and Bächi analyzed the PTT's integrated telecommunications system for the management and control of digital communications in the 1960s. Brugger examined how the IT project to introduce the new social security number in the 1970s changed organizational development within the federal administration. Koller examined the digitization of the federal administration using the example of ZAR, thus providing important preliminary work for this paper.[5]

Zur Soziologie einer Regierungstechnik, in: Lucien Criblez et al. (eds.), Staatlichkeit in der Schweiz: regieren und verwalten vor der neoliberalen Wende, Historische Bildungsforschung, Vol. 2, Zurich 2016, pp. 171–191; Kijan Espahangizi, Moritz Mähr, The Making of a Swiss Migration Regime: Electronic Data Infrastructures and Statistics in the Federal Administration, 1960s–1990s, in: Journal of Migration History, 2020, 6(3): 379–404; Moritz Mähr, Kijan Espahangizi, Computing Aliens. From Central Control to Migration Scenarios, 1960–1990, in: Monika Dommann et al. (eds.), Data Centers: Edges of a Wired Nation, Zurich 2020.

5 Jon Agar, The Government Machine: A Revolutionary History of the Computer, History of Computing, Cambridge, MA 2003; Julia Fleischhack, Eine Welt im Datenrausch. Computeranlagen und Datenmengen als gesellschaftliche Herausforderung in der Bundesrepublik Deutschland (1965–1975), Zurich 2016, p. 22; Nick Schwery, Die Maschine regieren. Computer und eidgenössische Bundesverwaltung, 1958–1965, Preprints zur Kulturgeschichte der Technik, 2018 (29), https://doi.org/10.3929/ethz-b-000243303 (3/2/2021); Daniela Zetti, Die Erschliessung der Rechenanlage. Computer im Postcheckdi-

The first section of this article presents the historical context, the key players in the federal administration, and their agenda. The second section examines the minutes of the Expert Commission for Statistics on Foreigners and the expectations of the various stakeholders regarding the information system. The third section analyzes the final report of the Expert Commission. In this context, focus is placed on questions that were not clarified during the negotiation process in the Expert Commission. The fourth section shows how the Federal Council took back control of the project in response to tensions between the federal administration and the cantons and municipalities. The fifth section outlines the course of the pilot phase and shows how considering technical and organizational artifacts underscores the interdependence of automation and federalism. The conclusion summarizes the argument and suggests further research questions.

Statistics for growth and against fear of over-foreignization

As early as the beginning of the 1960s, Max Holzer, director of the Swiss Federal Employment Office, was convinced that Swiss labor market policy had to undergo fundamental changes. The system of «guest workers» was long regarded as a guarantee of low wage costs and as an implicit economic buffer. If economic growth were to slow down, residence permits would not

enst, 1964–1974, in: Gisela Hürlimann et al. (eds.), Gesteuerte Gesellschaft – Orienter la société, Zurich 2009, pp. 88–101; David Gugerli, «Nicht überblickbare Möglichkeiten». Kommunikationstechnischer Wandel als Kollektiver Lernprozess, 1960–1985, Preprints Zur Kulturgeschichte Der Technik, 2001 (15), https://doi.org/10.3929/ethz-a-004254297 (3/2/2021); Beat Bächi, Kommunikationstechnologischer und sozialer Wandel. «Der schweizerische Weg zur digitalen Kommunikation» (1960–1985), Preprints zur Kulturgeschichte der Technik, 2002(16): 85, https://doi.org/10.3929/ethz-a-004465764 (3/2/2021); Jérôme Brugger, At the Dawn of Swiss E-Government: Planning and Use of a Unique Identifier in the Public Administration in the 1970s, in: Administration & Society, 2018, 50(9): 1319–1334; Guido Koller, The Central Register of Foreigners. A Short History of Early Digitisation in the Swiss Federal Administration, in: Media in Action, 2017(1): 81–92, https://www001.zimt.uni-siegen.de/ojs/index.php/mia/article/view/6 (3/2/2021).

be issued. This would reduce the supply of labor, which in turn would have a positive effect on the unemployment rate.[6]

The recruitment areas for cheap labor were increasingly extended to avoid rising wage claims in the countries of origin. While in the 1950s workers were mainly recruited from neighboring countries and southern Western Europe, in the 1960s Switzerland was also recruiting in Greece, Yugoslavia, and Turkey. At the same time, Italy was demanding better working conditions for its citizens and access to the Swiss social security system, permanent right of settlement, and simplified family reunification.[7]

Holzer assumed that at some point Switzerland would give in to these demands. The proportion of foreigners in the permanent resident population would then increase and would stabilize at a high level. Family reunification would lead to demographic shifts and make many new homes, schools, hospitals, and transportation services necessary. In order to study this scenario scientifically and to derive specific policies, Holzer suggested to the Department of Economic Affairs in 1961 that a study commission be set up. Holzer himself was appointed chairman of the subsequent commission to study «the problem of foreign workers».[8]

The government's issue with foreign workers was that future migration policy had to reconcile the conflicting claims of various political groups. On the one hand, the rapidly growing economy demanded more and more cheap labor. On the other hand, trade unions and national-conservative groups feared the economic consequences of migration – inflation, wage dumping, and housing shortages – and the rationally elusive political and

6 André Holenstein, Mitten in Europa: Verflechtung und Abgrenzung in der Schweizer Geschichte, Zurich 2014, pp. 307–328. A detailed description of the guest worker system can be found in Marcel Berlinghoff, Das Ende der «Gastarbeit»: europäische Anwerbestopps 1970–1974, Studien zur historischen Migrationsforschung (SHM), Vol. 27, Paderborn 2013, pp. 75–97.

7 Matthias Hirt, Die Schweizerische Bundesverwaltung im Umgang mit der Arbeitsmigration: sozial-, kultur-, und staatspolitische Aspekte von 1960 bis 1972, Saarbrücken 2009, pp. 64 ff., 219 ff.; Tobias Senn, Hochkonjunktur, «Überfremdung» und Föderalismus: kantonalisierte Schweizer Arbeitsmigrationspolitik am Beispiel Basel-Landschaft 1945–1975, Zurich 2017, pp. 27–40.

8 Ausländischen Arbeitskräfte.

demographic «danger of over-foreignization». For Holzer, it was obvious that immigration would have to be restricted in future. But to what extent immigration could be restricted and whether the «quality» of immigration could be controlled was unclear. The proportion of foreign workers in labor-intensive and low-productivity, low-wage sectors such as agriculture was very high and could be reduced by modernizing machinery without impairing economic growth. In industry and the service sector, however, the country was dependent on foreign skilled workers. The government thus had to find a way of restricting immigration so as to allow the labor market continued access to skilled workers from abroad without fueling the xenophobic resentment of the population.[9]

In its influential final report of 1964, which ran to nearly 300 pages, the study commission made various proposals for reforming labor market policy. The decentralized control of labor supply by the cantons was felt to be inefficient. For this reason, the commission argued, the Federal Council should intervene in the cantons' competencies to steer them toward a centralized allocation of work, residence, and settlement permits based on economic and demographic considerations. To this end, the commission also called for better social and labor market statistics. Only with complete, accurate, and regularly collected figures would it be possible to manage labor migration effectively and in a growth-friendly manner. Good figures were increasingly becoming an important tool for the federal administration; they promised legitimacy, credibility, and visibility. Statistics served «agenda setting» in the political debate.[10]

In late 1964, shortly before the report was published, the issue of labor migration became a hot topic among the public. Italy had renegotiated the 1948 «emigration agreement» with Switzerland and wrested some concessions from it. Swiss national-conservatives considered the government's concessions in the area of settlement rights and family reunification excessive.

9 Hablützel, Gilg, Beschleunigter Wandel; Damir Skenderovic, Gianni D'Amato, Mit dem Fremden politisieren: rechtspopulistische Parteien und Migrationspolitik in der Schweiz seit den 1960er Jahren, Zurich: Chronos 2008, pp. 31–68.

10 Bericht der Studienkommission für das Problem der ausländischen Arbeitskräfte, Bern 1964. Hans Ulrich Jost, Von Zahlen, Politik und Macht: Geschichte der schweizerischen Statistik, Zurich 2016, pp. 5–100.

The National Action against the Over-Foreignization of the People and Homeland, a right-wing party, began collecting signatures for a popular initiative to restrict immigration. The so-called over-foreignization initiative called for limiting the proportion of foreigners in the total population to 10%. Since almost 15% of the population were foreigners and the demand for foreign specialists was still growing, this was a radical proposal.[11]

The Federal Council was convinced that the consequences of the initiative would be detrimental to the economy and immediately produced a range of measures. At the end of 1964, the Federal Aliens Police published a set of social statistics called «over-foreignization statistics», in reference to the commonly used expression «over-foreignization of the labor market» described in contemporary economics. Residence and settlement permits were used to assess the proportion of foreigners in the resident population. In addition, the Aliens Police issued circulars to encourage the cantons and communes to standardize the permit and registration processes. The Federal Council attempted to reduce support for the over-foreignization initiative among Swiss citizens by means of a particularly high profile policy: in January 1965, it issued a general entry ban for foreigners who could not produce an employment contract.[12]

Like the policies of the Federal Council, the debates in parliament and in the public arena increasingly focused on the available figures. Calls for a quantitatively measurable restriction on immigration became louder. The differing survey bases led to the circulation of various, sometimes contradictory, numbers. This created confusion and led to criticism of the Employment Office's statistics. Consequently, the Department of Economic Affairs commissioned the Central Office for Organizational Issues (Organizational Staff) in April 1965 to review the Employment Office's statistics.[13]

The Organizational Staff was a small executive staff attached to the Federal Council that supervised the Federal Computing Center and coordinated all of the federal administration's automation projects. Both its director, Otto

11 Gees, Europäisierungsprozess, pp. 121–135.

12 Hirt, Schweizerische Bundesverwaltung, pp. 54–55.

13 Swiss Federal Archives (BAR), E4300C-01#1998/299#19*, Brief Overview of ZAR's History.

Hongler, a lecturer at the Institute of Business Administration at ETH Zurich and president of the Swiss Society for Rational Administration, and his deputy, Hans Kurt Oppliger, an economist and former employee of the Bull computer company, were advocates of the operations research method of military planning. This method assumes that the success of organizational changes can be measured and mathematically optimized.[14]

The Organizational Staff considered the current statistics to be insufficient and supported the plan to collect statistics on foreigners more frequently and more accurately. However, no one in the federal administration had any experience with computerized registers and automated statistics of this magnitude, which were intended for continuous operation. The previous major projects of the Computing Center – the population census, the agricultural business census, and the commercial business census – were one-off statistical analyses that were planned from scratch every ten years. In addition, the Organizational Staff were convinced that the legal basis was insufficient for the continuous collection of data. For this reason, they recommended in their report that a broad-based expert commission be set up with the participation of the municipalities and cantons.[15]

This the Department of Economic Affairs did not do. Instead, it commissioned an internal administrative working group consisting of members of the Statistical Office, the Employment Office, and the Aliens Police to propose ways of improving the statistics.

The new statistics run on a mainframe computer in Bern

In 1965, this administrative working group started its work under the leadership of Max Baltensperger, head of the section for social statistics at the Statistical Office. One year later the group presented their initial results. The

14 BAR, E6500-02#1986/114#315*, Expert Report on Improvements in the Government Activities and Administrative Management of the Federal Council.

15 BAR, E7170B#1977/67#368*, Minutes of the 1st Meeting of the Expert Commission for Statistics on Foreigners. E6502-02#2002/226#16*, The Initial Situation for the Replacement and Enlargement Problem. E6502-02#2002/226#16*, Operating History of the Computing Center from 1966.

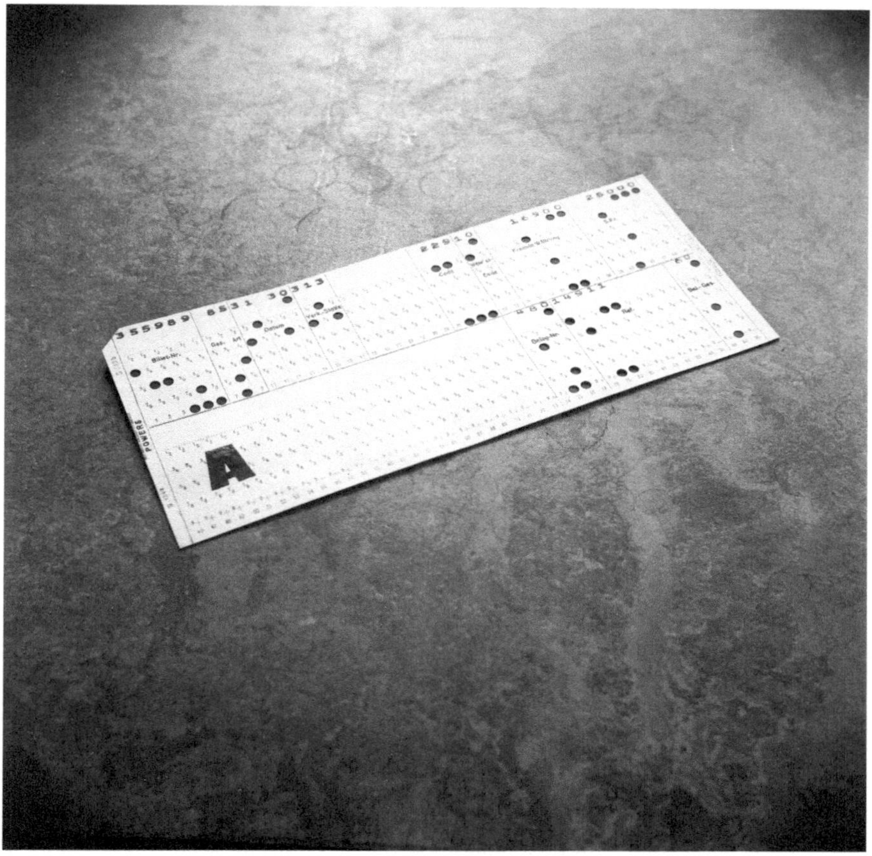

Fig. 1: Punched card (ETH Library Zurich, LBS_SR01-05328).

new statistics were to be run in the Computing Center on mainframe computers as a register of persons and to provide reliable figures at regular intervals. Note that, unlike today's personal computers, the mainframe computers available at the time were not operated interactively via screen, keyboard, and mouse. First, program instructions had to be punched onto cards (Fig. 1) and transferred to the processor unit (Fig. 2). Then data, also stored on punched cards or magnetic tape, was processed. Finally, the results were printed out or transferred to magnetic tape.[16]

16 BAR, Brief Overview.

Fig. 2: Front view of an IBM System/360 computer (ETH Library Zurich, Com_L19-0071-0002-0002).

The use of mainframe computers in private and public administration became popular in the 1960s, and statistical analysis was one of the main applications of these machines. IBM demonstrated the concept to the public with an art installation at the 1964 National Exhibition. Questionnaires were distributed in front of a sculpture characterized as a foreigner, and visitors were asked how much they identified with common stereotypes about the Swiss on a scale of 1 to 10. The machine-readable questionnaires of almost 600,000 visitors were read by a mainframe computer and compared with each other. Within a few moments, the computer printed out a result. Visitors could see for which questions and to what extent their answers deviated from the average.[17]

17 René Levy, Gulliver et la politique, in: L'Expo 64, Mémoire Vive, page de l'histoire lausannoise 9, Lausanne 2000; Koni Weber, Umstrittene Repräsentation der Schweiz: Soziologie, Politik und Kunst bei der Landesausstellung 1964, Historische Wissensforschung 1, Tübingen 2014.

The choice of technology seemed to be obvious for the working group: automated statistics on a mainframe. However, there was disagreement about who should run these statistics. No authority possessed the legal competence that would have made the decision obvious. Both the financing and the legal basis had to be clarified. The Statistical Office and the two closely cooperating authorities – the Employment Office and the Aliens Police – wanted to operate the expensive new statistics on foreigners. In order to clarify this issue, an expert commission was to be set up – as requested by the Organizational Staff a year earlier – with the participation of the municipalities and cantons.

The commission was set up by the Department of Economic Affairs on 24 May 1966. According to the official announcement, the group was tasked with clarifying whether the planned automation of statistics on foreigners could be implemented in practice. Under the leadership of Theo Keller, professor of economics at the University of St. Gallen, selected officials from statistical offices, employment offices, resident registration offices, and immigration police authorities from the cantons and municipalities examined the measures proposed by the working group.[18]

As an economist, Keller was an expert in quantitative methods. He had published on the advantages of administrative automation as early as the 1950s. At the end of 1965, as chairman of the Expert Group on Foreign Workers' Regulations, Keller had already drafted new restrictive measures for the Federal Council and had met the head of the Department of Economic Affairs, Hans Schaffner. In the following years he held key positions in several of the department's commissions.[19]

As director of the Employment Office, Holzer led the commission in a way that severely limited its scope of action. Over 20 participants attended commission meetings. Some participants who were also members of the administration's internal working group had already settled on which proposals should be more closely examined. Only two meetings were scheduled –

18 BAR, E7001C#1978/59#954*, Report of the Expert Commission for Statistics on Foreigners.

19 Theo Keller, Die wirtschaftliche Bedeutung der Automation, in: Schweizerische Zeitschrift für Volkswirtschaft und Statistik, 1958(I-3): 48–67.

one in August 1966 and one a few months later in November. Given the major financial and organizational implications of the proposals for municipalities, cantons, and the federal administration, this was a tight schedule.[20]

The first commission meeting was marked by sharp questioning and objections from the municipalities and the cantons. Adolf Ballmer, head of the Basel-Landschaft employment office and later president of the Association of Swiss Employment Offices, criticized the working group's proposals. He wondered whether a new method via the data center would be able to eliminate the main source of error, namely, the way the different communities collected their data. Some of them recorded the place of work, others the place of residence. Among other things, this practice would lead to outdated numbers or double counting. Ballmer spoke out in favor of harmonizing the cantons' data collection procedures. But he saw the decentralized federal structure of Switzerland's migration regime as an insurmountable hurdle.[21]

Marc Virot, head of the cantonal aliens police in Bern and president of the Association of Aliens Police, pointed out an important distinction: permits were counted, not foreigners. In most cases the data came from written reports from employers and landlords, not from the persons concerned themselves. In contrast to the authorities in the cantons and municipalities, the federal administration not only rarely came into direct contact with the persons affected but was also unaware of the consequences of its policies for the local authorities.[22]

Moreover, although the representatives of the cantons must have known that the Federal Council would be cutting back their powers even further in future, Virot could not imagine automating and standardizing the reporting system in view of the cantonal differences. It would certainly be easier for the central authority if the forms were uniformly designed. He did not believe, however, that the cantons could agree on uniform punched cards, as there

20 BAR, Minutes of the 1st Meeting; E7001C#1978/59#954*, Minutes of the 2nd Session of the Expert Commission for Statistics on Foreigners.

21 BAR, Minutes of the 1st Meeting.

22 Ibid.

were great differences, especially with regard to reported mutations (changes) in foreigners' records.[23]

The internal administrative working group reacted to these objections with slightly revised proposals, which they sent out only shortly before the second meeting. It was also decided to hold an extensive tour of the Computing Center before the second meeting in order to convince the commission members of the new technology. Subsequently, the statisticians took into account the concerns expressed at the first meeting. During the trial run, the existing statistics would certainly be continued. The switch to the new system would only be made when all concerns had been dispelled and flawless operation could be guaranteed.[24]

Commission chairman Keller also assured the group that the report was to be understood as a draft. The meeting dragged on until late afternoon. Provided that a trial run was carried out, the representatives of the cantons and municipalities agreed with the verdict that the new statistics on foreigners were technically possible. During the discussion, however, they noted that time would be short even if cantons and municipalities could be legally compelled to participate in the trial run. A working solution would surely take more time. Keller took a different view and announced that there would be no further meetings.[25]

As the minutes show, an extensive trial run was particularly a concern for the large municipalities and cantons, which had already automated their resident registration and aliens police authorities or were about to do so. They feared that a centralized registration process could lead to further shifts in competence toward the federal state. Before the report could be accepted by all members of the commission, a lengthy consultation process was necessary. Finally, the representatives of the cantons and municipalities expressed their support, as the final version of the report stated that various technical and organizational problems had yet to be clarified because the time available was too short or the analysis of the problems would be too costly. It was

23 Ibid.

24 BAR, Minutes of the 2nd Session.

25 Ibid.

also noted that reliable functioning of the statistics depended on thoroughly training the community.[26]

The promise of an automated migration policy was a rather vague and fragile compromise. Moreover, during the work of the commission the fronts had hardened. On one side were the administrators and planners of the federal administration who preferred a quick, technical solution to a political problem and were convinced of the top-down approach. On the other side were the representatives of the cantons and municipalities who could imagine neither that anything would change in the legal status quo – the extensive competences of the cantons – nor that a quick technical solution to the political migration problem was possible. The new statistics on foreigners depended on acceptance by local authorities and would only be successful if adapted to their processes and structures. However, little of this conflict was evident in the final report.

A final report, but still many questions

In March 1967 the report was distributed to all the authorities involved. The new statistics on foreigners were to combine the labor market statistics collected by the Employment Office since 1949 and the social statistics collected by the Aliens Police since 1964. None of the existing statistics represented the total number of foreigners living and working in Switzerland. The labor market statistics lacked non-employed foreigners. Social statistics did not include foreigners with short-term residence permits or cross-border commuters, who were not subject to mandatory checks. This gap in data collection would become even wider due to the new provisions of the migration agreement with Italy. The experts reckoned that in future more foreigners would have to be released from the obligation to undergo checks and that more non-employed foreigners would settle in Switzerland as a result of sim-

26 BAR, Minutes of the 1st Meeting; Minutes of the 2nd Session. Basel City already had a system in place in the mid-1970s that allowed recording of population mutations on screen and in real time. State Archives Basel-Stadt, FD-REG 8a 2–2 (1) 15/2, Project Group Population On-Line Mutations. BAR, Minutes of the 1st Meeting, First Draft; Report of the Expert Commission, Second Draft.

plified family reunification. The new statistics on foreigners should therefore cover all foreigners residing in Switzerland.[27]

The low collection rate of existing statistics was also seen as a problem by the Expert Commission. Since the 1960s, the Federal Council had been intervening more and more in cantonal labor market and admissions policy. The interventions were defined and implemented by the Employment Office. For this purpose, reliable and regularly collected statistics were needed. The new statistics on foreigners had to be updated on an ongoing basis. Ideally, at the end of every month, electronic processing would enable quick production of a detailed analysis.[28]

The structure of the new statistics on foreigners provided that all foreigners, i. e., all non-Swiss nationals with work, residence and settlement permits, be recorded in the system on a key date and the data stored on magnetic tape. In a next step, the data was to be compared with the data of the Social Security Register in Geneva and adjusted. Based on this «master tape», changes would be entered on an ongoing basis. In addition, statistical analyses tailored to «factual and temporal» needs were to be produced each month and shared with the cantons, municipalities, and federal authorities.[29]

This scenario was hypothetical for logistical reasons alone: the likelihood of being able to count one million foreigners or their permits on a specific date was at best illusory, even with the support of the cantons. Comparing this data against that of the Social Security Register was also wishful thinking. Although the Social Security Register wanted to acquire a computer for automatic processing, the planning was not very far advanced. This naturally left unanswered the question of how the data would be transferred from Bern to Geneva. Networking the two offices via a dedicated line was not even on the horizon. And transmission in paper form or by telephone was not only costly but also prone to errors.[30]

27 BAR, Report of the Expert Commission.

28 BAR, Report of the Expert Commission. Jost has examined in detail the prominent role of the Employment Office vis-à-vis Swiss economic and social statistics and describes it as autonomous and very close to the economy. Jost, Von Zahlen, Politik und Macht, pp. 94–97.

29 BAR, Report of the Expert Commission.

30 BAR, Minutes of the 2nd Session.

The choice of method for keeping the register, too, showed that the experts were unconcerned about data logistics. The «permanent inventory» method originated in accounting and required that all inventory data be kept centrally and on-site in Bern. In addition, new information or changes had to be continuously collected and transferred to the master tape every evening. This procedure was intended to guarantee that the data on the master tape was always up-to-date. It should also lead to a low error rate, as it was rare for two changes to be made to the same personal data record on the same day.[31]

Based on this daily updated master data set, the statistical analyses would then be carried out. The following questions would be answered daily: How many foreigners stayed or settled in the canton in a given year? How many of them were children? How many adults? What was the change in foreign population compared to the previous year? What were their countries of origin? How many foreigners with a residence permit were gainfully employed? How many foreigners with a settlement permit were gainfully employed? In which sectors were they employed? What professions did they pursue? These were mainly questions relating to the structure of the labor market.

The expert report also stipulated that entries and changes should be recorded in the municipalities and cantons and only be transmitted to the Computing Center in Bern for processing. It was argued that both the initial entry and the ongoing changes could be checked for accuracy more quickly locally, because it was often necessary to contact employers and landlords to verify the data. The short physical distance and the knowledge of local peculiarities seemed a compelling argument. The experts were also convinced that centralizing the collection of data would lead to an excessive increase in the number of staff in Bern.[32]

Decentralized data collection meant that the communes and cantons would not send forms with registration or change notifications to the Aliens Police. Instead, they would independently transfer the completed forms to an electronically processable medium. In this way, all registration and change

31 BAR, Report of the Expert Commission.

32 Ibid.

notifications would be collected in the respective offices and transferred to punched cards or magnetic tape. At the end of the month, these would be sent to the Computing Center in a collective consignment. The Computing Center would receive all entry and change notifications from the various offices and process them electronically on an ongoing basis. A few days later, stock lists would be printed out for each office and statistical analyses would be conducted. These analyses would then be made available to the offices by post. The report makes no mention of the frequency of analysis for the Employment Office or the Aliens Police.[33]

The report estimated the personnel requirement for the register at forty to sixty employees. About twenty of them were alphabet punchers. The total costs would amount to over two million Swiss francs. Even for a short report of 21 pages, these were extremely vague estimates of expenditure. Trained personnel in the field of electronic data processing were very rare and expensive. The experts were aware that in times of a shortage of skilled workers in the IT sector, it was unclear whether enough qualified personnel could be found. Moreover, the new statistics on foreigners would become one of the most expensive automation projects of the entire civil federal administration.[34]

The experts did not share the Organizational Staff's doubts about the legal foundation. They saw a sufficient basis for the first survey in the Statistics Act of 23 July 1870. For the operation of the register and the recording of current changes, the Aliens Police were given sufficient authority by Article 23, Paragraph 1 of the Federal Act of 26 March 1931 on the Residence and Settlement of Foreigners.[35]

The consensus expressed in the final report of the Expert Commission was relatively sparse. All members agreed that the current labor market and social statistics were inadequate and that better statistics should be produced as soon as possible. There was considerable disagreement on how such statistics and data collection should be implemented, apart from a few technical specifications such as centralized data management and the register manage-

33 Ibid.

34 Ibid.

35 Ibid.

ment method. The federal administration seemed confident that political pressure from the over-foreignization initiative would pave the way for a new technical solution. The cantons and municipalities seemed convinced that the legal status quo would remain intact and that the cantons could continue to implement their own entry policies. For this reason, the final report left important technical and organizational questions open, in particular questions of data logistics, which are central to an information infrastructure.

The Federal Council retakes control

The authorities involved were obviously keen to circulate the Expert Commission's report as soon as possible. Thanks in part to the popular initiative, for which the Federal Council was busy formulating a recommendation, the subject of over-foreignization was omnipresent. The Employment Office and the Aliens Police wished to exploit the momentum and urged the Federal Council to proceed with the new statistics on foreigners. It was imperative to launch a trial run of the new statistics on foreigners in a small area as soon as possible. The outcome should then serve as the basis for a decision on the overall project.

In the meantime, Elmar Mäder, director of the Aliens Police, applied for twelve full-time positions at the beginning of April to introduce the new statistics on foreigners. In the application he also mentioned that the circular letter of 1 January 1967 had already centralized the statistical recording of newly issued residence permits for non-active foreigners. For this policy, an application for an increase in personnel by two full-time positions had already been submitted in the middle of the previous year. In this way, the Aliens Police had moved forward and tried to set precedents for the institutional integration of the new statistics on foreigners.[36]

By then, all the agencies involved as well as the Federal Department of Justice and the Federal Finance Administration had weighed in on the expert report. With the exception of the Employment Office and the Aliens Police, all the authorities were critical. Precise deadlines and a detailed budget were lacking. In view of the high costs, a well-planned trial operation was essential

36 BAR, E6270B-01#1981/186#206*, Letter of 10 April 1967.

but could not be implemented in the foreseeable future. Too many technical and organizational questions were still unresolved. In addition to the resistance of other authorities, the over-foreignization initiative began to lose support, and the pressure to implement vigorous measures to restrict migration eased. On 29 June 1967, the Federal Council recommended that the initiative be rejected and announced a policy to restrict immigration. The new measures included a quota of foreign workers for each company that was calculated on the basis of previous years and could not be exceeded. New statistics on foreigners were unnecessary for these measures.[37]

The Aliens Police and the Employment Office wished to implement the new statistics on foreigners as quickly as possible and pushed ahead with their departments without authorization. Consequently, disagreement between the Department of Economic Affairs and the Department of Justice arose. In a strongly worded communication, Mäder and Holzer were urged to stick to official channels. No personnel budget would be discussed until further notice. This was a defeat for the Aliens Police and the Employment Office. The Statistical Office now profited from this discord. As it was neither part of the Department of Justice nor the Department of Economics, but rather the Department of the Interior, it was able to act as an independent and impartial party. The Federal Council instructed the Statistical Office to clarify the costs of the trial run and continuous operation as well as the open technical and organizational questions.

At the beginning of January 1968, the Statistical Office began to collect, prioritize, and process the questions that the expert report had left open. After the Aliens Police and the Employment Office had made many decisions on their own initiative, the process now returned to the official channels of the administration. Other authorities were involved from the outset, and the administrative procedures and protocols were followed. This was explicitly recorded in the minutes. Only when all points had been clarified should the

37 Amtsdruckschriften, Volksbegehren gegen die Überfremdung. Bericht des Bundesrates, Vol. 4, Bern 1967, pp. 529–546.

new working group prepare its report. The draft was to be submitted to all authorities involved.[38]

When the Federal Council implemented the already announced restrictive measures on 28 February, the concern about Switzerland's over-foreignization initiative seemed to fade again. A few days after this announcement, the National Action withdrew its popular initiative. But the calm was short-lived. There was disagreement within the National Action about how to proceed. On 15 May 1968, James Schwarzenbach, a member of the National Council representing the National Action, launched a second, more radical over-foreignization initiative called the Schwarzenbach initiative. In contrast to the first popular initiative, the new initiative contained no withdrawal clause. In addition, the wording was more precise to make it harder to interpret the constitutional article in an immigration-friendly manner. This initiative gave new momentum to the discussion on new statistics on foreigners.

In October 1968, the Statistical Office submitted the first preliminary report of just under four pages to the Federal Council. The responses were brief and diplomatic. It would be possible to compile the statistics required by the Aliens Police and the Employment Office. The most fundamental registration offices would be the resident registration offices in the municipalities. The Aliens Police would act as a central collection and control point, coordinating the exchange of data with the cantons and communes. The Computing Center would run the new information system, carrying out the analyses and taking over the punching and checking of the documents. Thus were the most important open questions answered. A time schedule for the trial run could be determined, and the costs would not exceed the estimates of the expert report.[39]

The Federal Council decided to continue its investigation of the Statistical Office. In only three months, the Statistical Office, the Employment Office, and the Aliens Police agreed on a compromise solution. In January 1969 they submitted their final report to the Federal Council. A Central Aliens

38 BAR, E3321-01#1985/36#47*, Memorandum to the Meeting of 15 January 1968 on the Procedure for Clarifying Open Questions in the Report of the Expert Commission for Statistics on Foreigners.

39 BAR, E3321-01#1985/36#47*, New Statistics on Foreigners.

Register was to be created (Fig. 3). The Aliens Police would be given the lead responsibility for this project. Together with the Employment Office and the Computing Center, the Aliens Police would prepare various interim reports for the Federal Council: the first after discussions with the cantons and communes, and the second after trial runs with selected cantons. Both reports were to be submitted within one year.[40]

With this proposal, the Federal Council defused the conflict that had arisen between the federal administration and the cantons and municipalities. A one-year trial run would be carried out. All authorities involved were to be given a say. Expenditures would be distributed as evenly as possible between the Computing Center, the Aliens Police, and the cantons. Responsibility for the project was to be transferred to the smallest authority within the federal administration. In addition, the cantons and municipalities were to be given a voice by being an integral part of the co-reporting procedure and the pilot phase.[41]

The report also contained a timetable that showed not only the project milestones but also an estimated distribution of expenditures over the coming years. The total amounts did not differ significantly from the estimates in previous reports, but the costs were broken down more precisely and compensation for the local authorities was defined.[42]

What the report did not contain, however, was a clarification of the pressing legal issues. The Statistical Office was still convinced that neither the Statistics Act nor the Federal Law on the Residence and Settlement of Foreigners provided a sufficient legal basis. In the report these concerns were reduced to a statement that the legal basis would be sufficient, provided the Federal Council adopted the information system by the end of January 1970. The question of which laws would have to be created and whether this would limit the competence of the cantons was not raised. The timetable for the information system seemed to have been adapted to the political situation.

40 BAR, E3325-02#2013/10#23*, Activity Reports of the Computing Center; BAR, E3321-01#1985/36#62*, Co-Reporting Procedure; E3321-01#1985/36#84*, Summary Final Report.

41 BAR, Summary Final Report.

42 Ibid.

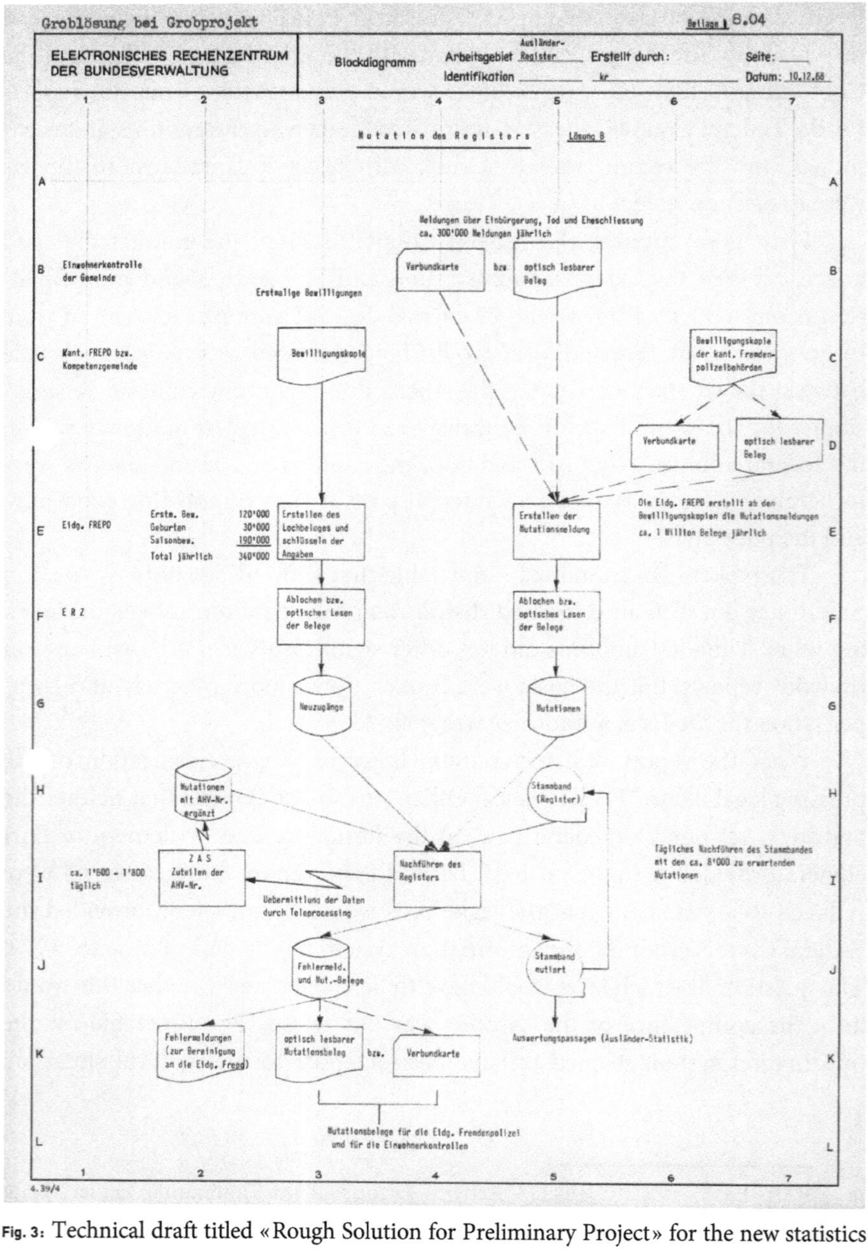

Fig. 3: Technical draft titled «Rough Solution for Preliminary Project» for the new statistics on foreigners (Swiss Federal Archives, E3325–02#2013/10#146*).

The second over-foreignization initiative, which aimed to radically limit migration, enjoyed a great deal of popular support. The Federal Council feared that approval of the initiative would mean major economic losses for Switzerland. The government was under increasing pressure.

Promise gives way to reality in the cantons

On 28 January 1970, the Federal Council commissioned the ZAR information system and, with a budget of over two million Swiss francs, approved the most expensive statistical information system to date. The mandate also included drawing up the missing legal basis for operating ZAR. In so doing, the Federal Council followed the final report of 1969 in all important points. The return to official channels and the political pressure exerted by the popular initiative had led the Federal Council to approve the trial operation. All on condition that the cantons and municipalities were involved in the implementation process from the outset.[43]

Meanwhile, the public debate continued to intensify in the context of the Schwarzenbach initiative. On 16 March 1970, the Federal Council responded to the pressure and tightened entry regulations for foreign workers throughout Switzerland. The nationwide restriction on immigration stipulated that foreign workers should be fairly distributed among the cantons according to economic and demographic needs. The labor market statistics of the Employment Office and the social statistics of the Aliens Police were used as the basis for calculating the distribution formula. In other words, the statistics were still compiled by hand at irregular intervals, a process that was too inaccurate and too slow in the opinion of everyone involved. The explanation was that the new statistics based on ZAR were still the subject of extensive planning meetings with cantons and municipalities.[44]

The negotiations with the cantons and the largest municipalities were a great challenge for a small federal authority like the Aliens Police. It did not have the necessary knowledge to negotiate the technical and organizational details of the upcoming trial run. A ZAR working group was set up to chan-

43 Ibid.

44 BAR, E3321-01#1985/36#103, Guidelines for Filling in Registration Forms.

nel the negotiations. This body, consisting of members of the Aliens Police, the Employment Office, the Statistical Office, and representatives of the cantons and communes, met at irregular but sometimes very short intervals. The group exchanged views on the work to be done and made technical and organizational decisions concerning the register.[45]

Most of the time in the meetings was spent resolving issues related to migration authority work in general, with little impact on ZAR: How should foreigners who left Switzerland or changed cantons without signing off be recorded? Should employers and landlords be obliged to report sign-offs to the authorities in addition to registration? How could small municipalities with part-time staff cope with the administrative burden? Could families continue to be registered using the family form, or did a separate form have to be completed for each person? Were the fees for the cantons and communes of 70 centimes for the initial registration and 20 centimes for the notification of a change appropriate?[46]

The discussions revolved around the everyday problems of the authorities on the front line. The only technical artifacts that were discussed very intensively were the new, uniformly designed, machine-readable forms for the initial and change notifications. The forms not only provided a semantic platform for the cantonal and municipal authorities to agree on common processes and interfaces but also represented the interface to ZAR as data supplier. Most of the authorities concerned would fill in the notifications by hand and send them by mail to the Aliens Police in Bern. Only a handful of authorities, mostly from large municipalities, expressed any interest at all in automated data transfer. The representatives of the Aliens Police were certain that authorities that already had an automated and centralized control system in place or were planning to do so would not cause any major problems in connecting to ZAR. Rather, the police worried about the smaller authorities, which processed everything by hand.[47]

The discussions took more time than expected, while the Federal Council pressed for the trial run to take place earlier than planned. Both the Aliens

45 BAR, E3321–01#1985/36#84, Report to the Federal Council of 3 December 1970.

46 Ibid.

47 Ibid.

Police and the Computing Center had overstretched themselves and were working at the limits of their capacity. The performance requirements for ZAR, as formulated a few years earlier by the Employment Office and the Aliens Police, had to be continually adjusted: registration of cross-border commuters, automatic assignment of AHV (old-age insurance system) numbers, and several tabular programs for special statistics were postponed to a later project phase. The goals of the federal authorities were pushed to the background. The success of the project depended solely on whether current and reliable figures from the cantons and municipalities could be compiled during the trial run.

The trial run itself was a balancing act. On the one hand, the technical feasibility had to be demonstrated; on the other hand, the new system had to be accepted by the cantons and municipalities. Without the full support of the cantons, it would not be possible to meet the schedule. Making the trial run as realistic as possible from a political and technical point of view, small as well as large, and manually as well as automatically, required the participation of the operating authorities. Various cantons made themselves available, and the final choice was made for Lucerne, Basel City, Grisons, and Neuchâtel. It was a balanced, Helvetic selection. Lucerne represented Catholic-conservative Central Switzerland, Basel was a large industrial city with a technically advanced population control system, Grisons was a large and rural canton from the south of Switzerland, and Neuchâtel represented the French-speaking part of the country.[48]

On 4 March 1971 the Aliens Police announced the ambitious timetable for the trial run. Within just one month, the four cantons were to clarify the organization and evaluate training materials for staff. From May to August, employees were to be trained in the municipalities. An inventory of all foreigners subject to controls in the canton of Lucerne was to be drawn up as early as 1 September. The other cantons would follow in the last quarter of the year. This meant stress not only for the test cantons but also for the Aliens Police and the Computing Center. The forms, envelopes, and training materials had to be printed in sufficient quantities after the final editing and distributed to the cantons. Adjustments to the forms or the survey process

48 Ibid.

always meant programming work for the Computing Center and could delay the trial by several weeks. In addition, the Aliens Police had to formulate and circulate the rules applicable to the cantons by circular letter.[49]

The Aliens Police quickly reached the limits of its capacity with the planning of the trial run. The situation came to a head because politicians and the Federal Council were pushing the pace. In order to avert acceptance of the Schwarzenbach initiative, effective policies had to be defined quickly and their effectiveness demonstrated with up-to-date figures. For these reasons, the technical issues took a back seat and the planning of the trial run focused primarily on migration police matters. These compromises served to water down the vision of an automated migration policy and signaled a return to the slow and incremental change that the federal system was accustomed to.

Conclusion

The planning of the ZAR information system was characterized as a negotiation process between the federal administration and the cantons and municipalities on the design of Switzerland's federal migration regime. The planning process for the information infrastructure and especially its technical artifacts provided a platform for presenting the expectations of the various stakeholders and agreeing on a consensus solution. This process did not proceed in a linear fashion, but in phases marked by sharp breaks.

In a first phase from 1964 to 1967, the federal administration dominated the discussion in the person of Employment Office director Max Holzer. The vision of automating migration policy appeared to Holzer to be feasible in view of the pressure from both domestic and foreign policy. In 1964 a new treaty with Italy committed Switzerland to unifying its migration regime in favor of foreign workers, and the National Action's xenophobic popular initiative led to a tightening of migration regulations. Holzer was convinced that sooner or later the cantons would forfeit their constitutional right to issue work, residence, and settlement permits. He wanted to anticipate this de-

49 BAR, E4300C-01#1998/299#31*, 1st Meeting with the Trial Cantons of 4 March 1971; E3321-01#1985/36#103*, Circular Letter from the Aliens Police of 1 July 1971.

velopment with a technical solution and set a precedent. His approach seemed to overestimate political pressure and underestimate the power of the federal status quo.

The solution proposed in the final report of 1967 had been worked out mainly on the drawing boards of the federal administration's statisticians. It required centralized structures and took little account of the realities of the federal periphery. The redesign of the federal migration regime was seen not as a political problem but as a technical one. The description of the new information system in administrative and technical terms made it difficult to involve the cantons and the municipalities in the planning process. The technical description deflected attention from the political consequences of this information system and the corresponding policies for the parties.

But precisely where the information system interfered with the processes and organizational structures of the authorities in the cantons and municipalities, there was great resistance to automation. The new migration policy that the technical design required was perceived by those involved as political rather than technical. In the eyes of the cantons and municipalities, too much power and data were concentrated in Bern. The new system was in conflict with the constitution, which placed work, residence, and settlement permits within the jurisdiction of the cantons. The Employment Office and the Aliens Police tried to ignore these concerns. Moreover, in order to reach a consensus as quickly as possible, important organizational and technical issues were excluded from the discussion and postponed until trial operation.

The indeterminate nature of the report and the haste with which the Aliens Police took action were the main reasons the Federal Council intervened in 1967 and transferred project management to the Statistical Office. The Statistical Office was part of the Federal Department of the Interior and thus independent of the Department of Economics and the Department of Justice. It concentrated on technical implementation and, in contrast to the Employment Office and the Aliens Police, was not involved in renegotiating Switzerland's federal migration regime at the policy level.

Thus, between 1967 and 1969 a solution was found which was also viable for the cantons. Not much remained of the visionary claim to automate migration policy. The information system was to replace the existing statistics without major adjustments in the cantons and municipalities. The operation of the register was to be entrusted to the smallest federal authority involved, the Aliens Police. In addition, the local authorities would be adequately compensated for their efforts and the federal administration would take care of local training. The demand for radical control of migration flows had given way to the federal reality in which cantons and the federal administration negotiated a compromise.

In 1970, the pressure on the Federal Council increased. The Schwarzenbach initiative, which would have capped the population share of foreigners at 10 % and would have forced hundreds of thousands of people to return to their home countries, found broad support among the voters and threatened to pass. On 28 January 1970, the Federal Council gave the green light for the new information system and demanded that the trial run be implemented sooner than planned. On 16 March, the Federal Council decided on a relatively radical, nationwide restriction on immigration. The policy was aimed at reassuring voters and preventing passage of the Schwarzenbach initiative. The policy required calculating the maximum number of aliens to be admitted on a monthly basis. These quotas could then be fairly divided among all the cantons. As long as ZAR was not yet fully operational, the old «overforeignization statistics» of the Aliens Police were to be used as a basis for calculating the quotas. Even before the trial run had begun, ZAR was already an integral part of an important policy. This also meant that any decision taken in the trial run would have direct consequences.

The trial run was planned for 1971. The time leading up to it was marked by further negotiations on the design of Switzerland's federal migration regime. The federal administration as well as the cantons and municipalities now had to agree on specific protocols and processes. In the discussions on data logistics and the interfaces to the ZAR information system, the basic features of an information infrastructure that would be in place until 2008 were defined. Many discussions revolved around technical artifacts such as machine-readable forms that could be filled out in pencil, and how to deal with inquiries about data. For most cantons and municipalities, forms were the most important interface for exchanging data, alongside the telephone. In 1974 the trial run was completed and ZAR was put into regular operation. This completed the planning but not the development of ZAR. As an information infrastructure that was in daily use by various authorities, it had to adapt not only to the changing policies of a highly dynamic federal migration regime but also to technological changes that shaped the organization and operation of the federal administration.

The planning of the information infrastructure for ZAR was not a linear process, nor was it done by the book. At no time were the requirements conclusively defined; the system design was developed only shortly before the trial run and had to be adapted several times during it. Rather, infrastructure planning for ZAR was a process of negotiation that engaged various actors over several years and was subject to the vagaries of politics and technological change. Various technical artifacts played a decisive role in this process. What was decisive was not the frictions and barriers created by the specific nature of ZAR. What was decisive was that the technical artifacts offered a semantic platform for the negotiation process. The forms, data logistics, processes, and protocols from the daily work of the migration authorities offered tangible subjects for discussion. In contrast, the promise of an automated migration regime, which was dominant in the early planning phase, was merely a project space filled with the desires of all involved parties.

Special measures: networking Swiss cantonal and federal universities

Daniela Zetti

Abstract

In the mid-1980s, the Swiss federal government launched special measures providing universities with funding for information technology and academic personnel to foster «education and further training as well as research in information technology for a period of five years». This article pursues the question of what made the special measures special across three fields of investigation. A first section shows that the federal government addressed the package to the socioeconomic challenges and dynamics of the era. A subsequent section discusses the special measures in line with mechanisms of cooperative federalism in science and university policy. The last section analyzes why a network for research and higher education (later called SWITCH) was well within the scope of federal science policy and discusses how it made a difference to university policy and Swiss federalism. The investigation is based on an analysis of resources from Swiss federal and university archives. It contributes to the history of federalism in the 20th century, to histories of computer science, and to the history of science and university policy.

Introduction

The Swiss network for research and education was planned in the mid-1980s as part of a package of «special measures» initiated by the federal government. These measures provided federal grants for investments in information technology (IT) for research and higher education for a period of five years. The term *special measures* points to the fact that they were exceptional. In the text that follows, I wish to examine the early history of the special measures and the network by asking what this «specialty» consisted of. The paper discusses why actors from the federal government and from academic institutions founded a scientific network. From 1986 until today the network is still known by the name SWITCH. Currently SWITCH is best known to users as an acronym and logo that can be found on login screens for academic email, library, and platform accounts. In the period investigated here the network was still without a name. Its planners had in mind to integrate academic services and to standardize access, but it would take time to realize these projects.[1]

From the very beginning, the network was intended to include and coordinate all of Switzerland's cantonal and federal universities. The institutions would be equipped with a great number of then brand-new computers called workstations to complement their already multifaceted fleet of digital machines. Sixty-two million Swiss francs was earmarked for the acquisition of the workstations. An additional 40 million francs would go for the purchase of a national supercomputer, and 15 million francs for the network. Most importantly, another huge part of the total of 207 million francs was reserved for temporary employment of personnel in computer science departments at universities and higher technical institutes. The official title of the document announcing the federal special measures – «Dispatch on special measures in favor of education and further training as well as research in information technology and engineering sciences» – clearly put the accent on education.[2]

As I began working on this paper, my investigative interest was twofold. First, I wished to explore the history of a computer network that was not yet operational during the period under study. It was still in the planning phase. A singular feature of the networking of Swiss universities is that it pioneered no new standards. It did not happen earlier than elsewhere (nor much later than in surrounding countries). No technology was invented, and there were no discussions or disputes about what standards or equipment should be used. Much was already in place, both in Switzerland and elsewhere. From a history of technology point of view, the fact that, indeed, very little happens in what follows is therefore due to the choice of the period under study. Nevertheless, I considered the early history of the network worth examining be-

1 This article is based on research led by a joint ETH Zurich–USI Lugano team on the early history of supercomputing in Manno and the SWITCH network (funded by the Swiss National Science Foundation, http://p3.snf.ch/Project-183007). It also benefits from insights from the «Trading Zones» research project and from a collaboration between the Chair for the History of Technology, ETH Zurich, and the Swiss Federal Archives. I would like to thank Henrike Hoffmann for acquiring further published and unpublished material, as well as for providing support for interviews at the University of Geneva.

2 Swiss Federal Gazette, 1986, 1: 321–383, on p. 374, table 6.

cause I wanted to know what prompted the special measures and how they differed from other measures to promote research and development at universities. That was my second research interest.

Maybe it's because I see the world through a particular lens – that of a historian of technology – but eventually I did stumble on a story of innovation. The planning and development of a network that would consist of components that both already existed but that had to be adapted and ordered abroad required technical, social, and organizational cooperation. That implied innovativeness. This initial finding squares with research results regarding the implementation of networks and development of digital standards in telecommunications in other countries.[3] More unconventional is my message about the planning methods. In history, planning is often synonymous with the postwar era, the associated years of prosperity, and unlimited belief in progress. Yet, in the last decades of the 20th century, planning methods continued to be refined and adapted to new missions.[4] What does that mean for

3 For a retrospective look at the development of packet switching, a fundamental technique for digital data transmission, during the period under review see Tony Rybczynski, Commercialization of Packet Switching (1975–1985): A Canadian Perspective, in: IEEE Communications Magazine, 2009, 12: 26–31. For pioneering Swiss research in packet switching, see the early work of Walter Neu and Albert Kündig, Project for a Digital Telephone Network, in: IEEE Transactions on Communication Technology, 1968, 16(5): 633–648. In the early 1980s, Swiss universities prepared local academic networks for research and teaching. For network planning at the University of Geneva within its setting of local multinational institutions, see University of Geneva, Archive (UniGE), rectorat, commission informatique. For a contemporary introduction to the concept of X.400 and the Swiss network's first standards and features, see Bernhard Plattner et al., X.400, elektronische Post und Datenkommunikation, Bonn 1993. On the politics of standardization of information networks, see the work of Andrew L. Russel, for example: Open Standards and the Digital Age: History, Ideology, and Networks, New York 2014. JoAnne Yates and Craig N. Murphy's «Engineering Rules» emphasizes that «computer networking usher[ed in] a new era in voluntary standard setting» – a field that was «well established and reasonably stable» in the 1980s but slow to react to new technologies. Joanne Yates, Craig N. Murphy, Engineering Rules: Global Standard Setting since 1880, Baltimore 2019, pp. 241–268.

4 On today's «thinking infrastructures» that configure entities, organize knowledge, and sort things out via rankings and ratings, see the case studies in the volume of the

the history of federalism after the 1970s? What does it mean for understanding the nature of a computer scientist's work? The concluding sections lay out a few hypotheses concerning developments in the field of science and university policy.

The aim of this article is to pursue the question of what made the Swiss special measures special across three fields of investigation. In a first section I will show that the first federal package of special measures was launched against a backdrop of structural change. The second section outlines the position of Swiss university policy and planning within the federal political system, especially in the context of a much-debated «cooperative federalism» in science policy. The final section analyzes events, debates, and studies that – within an institution dedicated to cooperative federalism – produced the first concept for a Swiss network for research and education.

The article is based on the research literature on Swiss science and university policy and on an analysis of source material from the Swiss Federal Archives in Bern, the ETH Archives in Zurich, and the Archive of the University of Zurich. An interview with Jürgen Harms of the University of Geneva gave me incredibly valuable advice on the background, scope, and goals of the Swiss network for research and education.[5]

Federal special measures in response to socioeconomic problems

Today the term *special measure* is synonymous with «a need for haste» in applying for potential funding, as a small window of opportunity is immediately going to close again.[6] The concept appeared prominently for the first

same name edited by Martin Kornberger et al., Thinking Infrastructures: Research in the Sociology of Organizations, Vol. 62, Bingley 2019.

5 The original language of sources is German and French. Quotes are translated for this publication.

6 An ad from 2011 gives the following instruction: «Knowledge and technology transfer is to receive additional funding of over 200 million Swiss francs. Applications can be submitted until mid-December. Nevertheless, there is a need for haste, as applications will

time in Swiss federal parliamentarism with the issuing of the government's dispatch. The implication was a one-time chance to get something funded and done quickly, but other special measures were launched later. The first *Informatikpaket* (or *paquet* in French) was followed by a second one (1986–1991), and by even more special measures passed by parliament up to the early 21st century. Subsequent measures were concerned with the twin issues of continuing education and IT. In other words, for more than a decade the federal government tried to steer development that was perceived to be triggered and characterized by the dynamics of «structural change». Along with «digitization» of society, this structural change stimulated investments in lifelong learning and continuing education.[7]

In the mid-1980s, structural change was perceived as an urgent but long-standing issue, especially in the light of other urgent problems. Many burning issues of the decade were easier to tackle. In 1985, the Federal Council, as the Swiss government, stated that «assessment of the content of the first half of the legislative term showed that significant deviations from the government program had become necessary, especially in three areas of responsibility». The first area was migration policy – the government was facing a «sharp increase in asylum applications».[8] Second, there was a «worry-

be dealt with on a first-come, first-served basis», www.startupticker.ch/en/news/september-2011/sondermassnahmen-des-bundes–jetzt-gesuche-einreic (12/2/2021).

7 A brief outline of federal measures to promote continuing education (*Weiterbildung*) in the field of technology since 1978 is given by Michael Geiss: «Microelectronics, in particular, served as the driver of an accelerated Swiss continuing education policy.» Michael Geiss, Sanfter Etatismus, Weiterbildungspolitik in der Schweiz, in: Lucien Criblez et al. (eds.), Staatlichkeit in der Schweiz. Regieren und Verwalten vor der neoliberalen Wende, Zurich 2016, pp. 219–246, on p. 234. Tobias Straumann deals with yet another contemporary federal package of measures that was supposed to combine regional and technology policy but was rejected by a referendum in 1985: Tobias Straumann, Rezession, Technologiepolitik und Risikokapital. Das Scheitern der Innovationsrisikogarantie 1985, in: Schweizerische Gesellschaft für Wirtschafts- und Sozialgeschichte / Société Suisse d'Histoire Economique et Sociale, 2001, 17, 403–419.

8 Report on the management of the Federal Council, the Federal Supreme Court, and the Federal Insurance Court in 1985. Federal Council Annual Reports, 1985, 120: 3. On

ing development of forest dieback».[9] The government made another entry to the files of the federal administration: «Finally, the rapid structural change in the economy, which we have already highlighted in the government program, has led us to propose a number of additional measures for funding.»[10] Directly following this passage, the Federal Council explicitly mentioned the special measures to promote IT and engineering.[11]

In launching the special measures, Swiss politicians were thus reacting to perceptions that were common in the Western world.[12] The special measures simultaneously reflected challenges confronting the Swiss economy in a period that, according to contemporary observers, heralded the decline of the industrial sector and the rise of the service sector.[13] However, economic historian Jakob Tanner has pointed out that this popular description of sectorial change overlooks something, namely,

---

planning efforts in federal migration policy, see Moritz Mähr's contribution in this special issue.

9 Rapidly initiated protective measures, such as the introduction of catalytic converters in cars, successfully halted the forest dieback predicted in the 1980s. Monika Gisler emphasizes that the federal government's measures were important for adapting the universities' research policies to environmental problems. Monika Gisler, Wie die Umwelt an die ETH kam. Eine Sozialgeschichte der Umweltnaturwissenschaften, Zurich 2020, p. 20.

10 Federal Council Annual Reports, 1985, 120: 3.

11 See Alban Frei's dissertation for an overview of the research literature on recent Swiss economic history and its resonances in science policy and management: Alban Frei, Sichtbare Netzwerke. Forschungspolitik und Life-Sciences zwischen 1990 und 2016 in der Schweiz. Eine Fallstudie zu SystemsX.ch, Zurich 2018.

12 Lutz Raphael, Jenseits von Kohle und Stahl. Eine Gesellschaftsgeschichte Westeuropas nach dem Boom. Berlin 2019. The Swiss PTT's (postal services, telegraphy, and telephony) archive gives an impressive insight into «postal change» during the liberalization and partial privatization of the former state-owned PTT enterprise: https://www.oralhistory-pttarchiv.ch/de/themes/liberalisierung-post (11/2/2021).

13 Sociologists Alain Touraine and Daniel Bell coined the term *post-industrial society*. Alain Touraine, La société post-industrielle. Naissance d'une société, Paris 1969. Daniel Bell, The Coming of Post-Industrial Society: A Venture of Social Forecasting, New York 1973.

> the fact of an interpenetration of industrial goods production and service provision that undermines the three-sector distinction introduced after World War II. Especially in the watch industry, which was able to stabilize at a considerable level after a crisis-like adjustment shock, this service intensity of goods production and, conversely, the dependence of most services on industrial goods (such as computers) became apparent.[14]

The interconnectedness of IT services and production described by Tanner suggests that structural change meant changes for companies and job markets, which were linked to each other through various dependencies. That, in turn, required the actors to be able to trust operational, communicative, and administrative structures. In the eyes of contemporaries, the announcement of comprehensive change made it obvious that the federal state – not individual cantons or even the private sector – had the say to act and to invest. In emphasizing advanced technology and education, the federal dispatch was asserting that parliament and the institutions of the federal state were in charge of managing change.

Apart from honing in on the era's socioeconomic challenges and dynamics, what was it that made the dispatch «special»? What ramifications did it have for science and university policy? And in what ways did introducing the measures help to transform the federal state? Answering these questions requires a look at the *courant normal* of Swiss science and university policy since the 1960s.

Ascent and crisis of cooperative federalism in university policy

Over the last half-century, Swiss science policy has become increasingly federal. In the words of Sebastian Brändli, during the 1960s there was finally movement in Swiss higher education policy thanks to the federal govern-

14 Jakob Tanner, Die Geschichte der Schweiz im 20. Jahrhundert, Munich 2015. On the crisis related to the reforms of the world monetary system and on the decline of the gross domestic product in the years from 1974 to 1976, see pp. 415–420. For the diminishing importance of the industrial-commercial sector in the 1970s and its qualitative knowledge-based reorganization in the 1980s, see pp. 483–485, here p. 485.

ment's use of «the constitutional provision of 1874 (!), according to which the federal state could subsidize the cantonal universities».[15] The role of the federal state as well as its relations with the cantons is thus an important topic in the research literature on Swiss science policy. In the 1960s, the protagonists and institutions of the federal state consequently recognized the importance of science and higher education policy.[16] The federal government commissioned several expert reports on higher education. In 1965, the Federal Council installed a body of experts – the Swiss Science Council – to advise it on university matters.[17] In 1969, the federal state took over the École Polytechnique de l'Université de Lausanne, which was henceforth called École Polytechnique Fédéral de Lausanne (EPFL). The polytechnic became a sister institute of the older federal institute of technology, ETH. ETH is based in German-speaking Zurich. Ever since its founding in 1855 it had been one of the federal state's most important investments in higher education.[18]

Over a hundred years after its first initiative in the field of higher education, the national legislature laid the groundwork for providing financial support to the Swiss cantonal universities. In June 1968, the Federal Assembly passed the Federal University Support Act (*Hochschulförderungsgesetz*; *L'aide aux universités*). The law regulated the distribution of financial resources to the cantonal universities. It envisaged the creation of a body that

15 Sebastian Brändli, Investition Bildung. Ausbaupläne und Reformpläne der Zürcher Universitätspolitik vor 1968, in: Erika Hebeisen et al., Reformen jenseits der Revolte. Zürich in den langen Sechzigern, Zurich 2018, pp. 11–25, on p. 12.

16 On Swiss science policy with an emphasis on its transitive and protean character in the middle of the 20th century, see Frédéric Joye-Cagnard, La construction de la politique de la science en Suisse. Enjeux scientifiques, stratégiques et politiques (1944–1974), Neuchâtel 2010. On federal science policy in a national and international «multilevel reality» (1960–1990), see Thomas Gees, Viel Diskurs – wenig Steuerung. Schweizer Wissenschaftspolitik in der Mehrebenenrealität, in: Criblez et al., Staatlichkeit in der Schweiz, p. 318: «That Switzerland needed a science policy was undisputed among the various players; the debate turned [...] mainly on the level at which it should be controlled.»

17 Schweizerischer Wissenschaftsrat / Conseil suisse de la science / Consiglio svizzero della scienza.

18 David Gugerli et al., Transforming the Future: ETH Zurich and the Construction of Modern Switzerland 1855–2005, Zurich 2010.

would promote, even require, the exchange of information between the cantonal universities and the federal state.[19]

The University Support Act and the soon to be founded Swiss University Conference are today considered the embodiment of «cooperative federalism», which represents a revival of relationships between the cantons and the federal state in the 1960s. Historian of education Marc Herren has pointed out that cooperative federalism implies cooperation of various kinds:

> The term «cooperative federalism» is used to describe different relationships and forms of cooperation between the member states and between the member states and the federal state. In particular, a distinction is made between horizontal and vertical cooperative federalism. In the first case, only the constituent states are involved; in the second case, both member states and the federal state are involved. There are also mixed forms, such as when the federal state is involved but does not have a voice.[20]

Universities were located in eight out of 25 cantons of Switzerland. Entire regions and the whole of Italian-speaking Switzerland had no university. Nevertheless, reading the University Support Act in its 1968 version, it is clear that it embodies the spirit of cooperative federalism. Consequently, cooperative federalism experienced a boost in the field of higher education pol-

---

19 Federal University Support Act [414.20], in: Swiss Federal Gazette, 1968, II: 10–21, https://www.amtsdruckschriften.bar.admin.ch/viewOrigDoc.do?id=10044039 (11/1/2021); Gees, Viel Diskurs, p. 321. Marc Herren, Die Nationale Hochschul- und Forschungspolitik in den 1960er und 1970er Jahren, in: Lucien Criblez (ed.): Bildungsraum Schweiz. Historische Entwicklung und aktuelle Herausforderung, Bern 2008, pp. 219–250, on pp. 222, 229.

20 Herren, Forschungspolitik, p. 221. Dietmar Braun prefers to call Switzerland's federalism a «decentralized» federalism: «At least on the revenue side, Switzerland is not yet concerned with ‹cooperation› […]. It is only on the expenditure side that the interdependent network between the federal state and the cantons occurs in Switzerland as well, justifying the view that Switzerland is also a ‹cooperative federalism,› i.e., a closely intertwined system.» Dietmar Braun, Dezentraler und unitarischer Föderalismus. Die Schweiz und Deutschland im Vergleich, in: Swiss Political Science Review, 2003, 9: 57–89, on p. 63.

icy.[21] What kind of exchange between the Confederation and the cantons was envisaged? The federal government demanded information exchange within and between the universities. The tasks of the future Swiss University Commission therefore included «maintaining contact with the student body» and «issuing guidelines on admission requirements, curricula, and examination regulations for the various fields of study, as well as on the mutual recognition of examinations and degrees». Both the University Conference and the Science Council, however, drew up recommendations only; the federal state was not authorized to speak or implement laws. From a legislative and judicial standpoint, these were low-key interventions by the nation-state.[22]

In return for information exchange, the act allowed the cantons to receive financial support from the federal state – in the form of fixed budgets for operating costs (excluding funds for personnel) and funding for urgent structural investments, especially for buildings. The recipients of the funds were the cantonal universities. Another task of the University Conference was to «examine applications for grants for material investments». The universities had to report to the University Conference on the use of the funds.[23]

The establishment of EPFL took place in parallel with these measures. It emphasized a well-known division of labor that had existed between the federal state and the cantons since the founding of the federal state in the middle of the 19th century: the Confederation was responsible for polytechnic schools, and the cantons for universities.[24] However, the University Support

---

21 The same holds true for German federalism of the same era. Education policy is an important field of federalism. Dieter Langewiesche writes with a view to today's Germany: «No matter how one judges the role of the member states [...] they are deeply anchored in the political-social order. In school and education policy, they continue to be the main actors and have created their own instances of coordination.» Dieter Langewiesche, Vom vielstaatlichen Reich zum föderativen Bundesstaat. Eine andere deutsche Geschichte, Stuttgart 2020, p. 105 f.

22 Federal University Support Act, p. 17. Since the canton of Jura was created in 1979, Switzerland consists of 26 cantons.

23 Ibid.

24 Around 1900 the Swiss national state was hardly interested in scientific research. «It was only in the three decades of the middle of the 20th century that mechanisms for a

Act, the expert commissions, and the Science Council reinforced a new kind of involvement of the nation-state. In the eyes of the cantons, it became involved in the field of university planning by demanding reporting and «coordination» among its constituent entities. For contemporaries, the demands of the federal state were double-edged. Thomas Gees has emphasized that during a period of strong educational expansion, the cantons were highly appreciative of the federal state as a funder, but at the same time were wary of national concepts. Marc Herren underscores how careful the federation had to be when evoking coordination and cooperation. An impression quickly formed that coordination was a matter of duty and synonymous with streamlining.[25] Summarizing the federal state's efforts in the field of science policy in the 1960s, Frédéric Joye-Cagnard has this to say: «[O]n the eve of the 1970s, the Confederation set up a set of founding bodies for its science policy, without, however, creating a federal department entirely dedicated to this activity.»[26]

Moreover, as far as the area of university funding is concerned, the federal state narrowly failed to establish a federal university. Just as striking as the multiformity of mechanisms and multiplicity of actors in contemporary Swiss research and higher education policy is the federal state's emphasis on material investments in universities and formal reporting on the need for funding. In other words, it did not build its own university, but rather provided the material to enlarge the existing cantonal universities and influenced the way the universities were managed. Looking back self-critically in 1983, the Federal Council noted that in 1968 it had clearly overestimated the importance of presumably one-off building investments compared to the

science policy began to be put in place.» Joye-Cagnard, La construction, p. 15. For ETH Zurich, the founding of EPFL had far-reaching consequences. Among other things, its supervisory authority, the «School Board» (*Schulrat*), was transferred to a new institution, the «ETH Board», which since then has supervised all federal institutions of higher education and research (the so-called ETH Domain). Gugerli et al., Transforming the Future.

25 Gees, Viel Diskurs, pp. 318, 320. Herren, Forschungspolitik, pp. 228–232.

26 Joye-Cagnard, La construction, p. 352.

universities' operating costs. The latter in particular had been rising continuously for 15 years.[27]

At the University of Zurich, the internal changes initiated in order to comply with the University Support Act date back to the early 1970s. Zurich serves as an example of a cantonal university that reacted quickly to the new law. The university's annual report for 1972 testifies to adjustments and even an attempt to anticipate further legislation: «The core of a planning organization was created [...] in the form of the Planning Commission and a Planning Committee, which will ultimately deal with all matters of medium-term university planning – not least in view of the requirements that will be imposed by today's or a future federal University Funding Act.»[28] The university also created a vice rectorate responsible for planning, led by business economist and longstanding head of the Planning Commission, Edwin Rühli. Rühli's approach to planning was well integrated into his daily work as an academic. For example, he taught lectures within the university curriculum on leadership tasks and university planning. The Planning Commission published 15 extensive studies in the period from 1973 to 1997.[29]

Archivist Inge Moser recently recorded the planners' inventory for the university's archive and described how closely planning and computing were connected:

> The institutionalization of planning and the increased surveying of the university since the 1970s did not just happen by chance simultaneously with the development of information systems and computer sciences. There was close collaboration with the Computer Center and computer scientists, and the planning staff often partici-

27 Dispatch on loans for the fifth funding period under the University Funding Act, in: Swiss Federal Gazette, 1983, 2: 221–270, on p. 224. In the 21st century, the aesthetics of concrete are characteristic even of data centers. See Monika Dommann et al. (eds.), Data Centers: Edges of a Wired Nation, Zurich 2020.

28 Annual Report University of Zurich 1973, p. 16. www.archiv.uzh.ch/de/editionen/jahresberichte.html (1/10/2019)

29 Inge Moser, Hochschulplanung an der Universität Zürich in den 1970er Jahren, www.archiv.uzh.ch/de/vitrine/aeltere_beitraege.html#29 (1/10/2019). A look into the files of the archives of the ETH Board shows that the management of IT for Lausanne and Zurich has preoccupied the council since the 1970s. IT in administration and research was an overarching issue.

> pated in projects to introduce documentation systems and databases, which in turn would serve as tools for planning.[30]

Up to 1981 the university had been busy building an entirely new and second campus at the Strickhof site just outside the city center.[31] In 1986 Rühli stated that in analyzing data, he had identified «three main conclusions from the planning: the divergence between the university's mission and its resources [was] steadily increasing; the explosive growth in computing require[d] special efforts; and a pronounced need for replacement [was] emerging in the area of apparatus.»[32] The university had diagnosed itself as having aging equipment that perhaps needed to be upgraded, or perhaps replaced with IT. Yet, through it all, one problem remained: even with the help of professional and computer-aided planning, the university had not come to grips with problems of growth and rising costs. This was a stinging indictment of the phenomenon known as structural change.

The frustration was palpable: planning efforts had failed to eliminate the financial problems that had arisen for the universities because of the steady and striking growth in student numbers. Additional, pressing equipment problems were added to the mix. And the same issues manifested elsewhere. By the mid-1980s, planners had a poor image among academic staff. The problem is well illustrated in a talk given by Urs Hochstrasser, director of the Federal Office for Education and Science and an outspoken friend of plann-

30 Moser, Hochschulplanung, www.archiv.uzh.ch/de/vitrine/aeltere_beitraege.html#29 (1/10/2019).

31 Sebastian Brändli points to the fact that «an important factor in rising student numbers was the steadily increasing proportion of female students». Brändli, Investition Bildung, p. 15.

32 «Prof. Rühli nennt drei Hauptschlüsse aus der Planung: Die Divergenz zwischen dem Auftrag der Universität und ihren Mitteln nimmt stetig zu; das explosive Wachstum im Bereich Informatik verlangt besondere Anstrengungen; bei den Apparaten zeichnet sich ein ausgeprägter Ersatzbedarf ab.» Archive of the University of Zurich (UAZ), F.3.110 Prorectorate Planning, Development Plan 1986–1991: General Files 1984–1986 (PSA 501 19 (1)), Current Planning.

ing,[33] to the Association of Swiss University Teachers titled «University planning from the federal government's perspective».[34] Hochstrasser was obviously at pains to explain to his audience what the workings and goals of federal higher education funding were and why it was important that the audience support planning. In speaking of the need for the Confederation to be frugal in all its expenditures, he addressed those present as citizens. He asked for understanding of the view of federal policymakers who did not wish to support universities with fixed budgets or to be fettered by long-term planning. «To maintain adaptability to unexpected new demands in research and teaching, it is necessary that donors not insist on stubborn adherence to plans but also be prepared to take special measures on short notice to cope with exceptional situations.» Hochstrasser spoke of an «emergency» that had arisen in the field of computer science and that threatened the «prosperous development of our industrial state».[35]

His statements showed how delicate his mission was: «A prerequisite for such extraordinary efforts is, of course, the optimal, full utilization of normal federal aid, which in turn is not possible without good planning.» Hochstrasser had to perform a balancing act in front of the university professors: planning wasn't everything, but it was very important. «Planning wasn't everything» meant that federal policymakers were allowed, beyond the agreed-upon, planning-based funding of the University Funding Act, to set substantive and extraordinary priorities to «push through important higher education policy goals». At the same time, planning remained important: «With all due respect for the reservations of university members about excessive planning, the federal authorities [...] nevertheless hope that the faculty of our universities will continue to support our common planning concerns in the future.»[36]

33 On Hochstrasser, see Andrea Weibel, «Hochstrasser, Urs», in: Historisches Lexikon der Schweiz (HLS), Version of 23/5/2005, https://hls-dhs-dss.ch/de/articles/024735/2005-05-23/ (5/3/2021).

34 Urs Hochstrasser, Hochschulplanung aus der Sicht des Bundes, in: Bulletin der Vereinigung Schweizerischer Hochschuldozenten, 1985, 1: 18–24.

35 Ibid., pp. 21, 23.

36 Ibid., p. 24.

Hochstrasser's oration provides another hint of what made the special measures special. First, the federal dispatch defined measures that were targeted to investments in IT and personnel, not building materials. Second, the measures were intended to be seen as a complement to the University Funding Act and its disputed administrative routines. This perspective was risky. The special measures might well trigger a crisis of cooperative federalism because they were grist to the mill for people fed up with federal monitoring and control. Moreover, in affecting the universities' «content control function» (*inhaltliche Steuerungsfunktion*), the measures reinforced well-known problems of cooperative federalism. In 1989, the cantonal Education Directorate of Zurich let its university know that it «ha[d] always been fundamentally opposed to ‹special funding›», while it did tolerate specific project-based funding by the Swiss National Science Foundation. The Directorate listed three reasons for its opposition to special measures:

> [F]irst, such funds [...] are taken away from regular funding and, second, the Confederation in effect takes over the content control function, which it is not constitutionally entitled to do. Finally, it is by no means desirable to have the federal state get its hands on areas that have become competitive through cantonal efforts (neuroinformatics in Zurich, for example).[37]

The Zurich Education Directorate was convinced that the measures did not stand to increase the bottom line. They represented illicit interference into the cantons' autonomy. On top of that, the confederation would reap the fruits that the cantons had sown. The directorate concluded that in the future, the federal state should use special measures only to fund continuing education, not university research and teaching. «The special federal funding that has now been decided should be limited to overarching areas, such as further education; the specific funding of individual areas is a matter for the Swiss National Science Foundation.»[38] In the same breath, the directorate

37 UAZ F.3.116, Rectorate, Development Plan of the University of Zurich 1990–1995, Preliminary Discussions with the [Cantonal] Higher Authorities Education Directorate, Finance Directorate, Health Directorate, Construction Directorate, 10–14 April 1989, p. 7.

38 Ibid.

linked special measures to «research projects». It expressed its hope and desire that the latter would involve «specific» funding and «individual» areas. Projects required grant applications, which in Switzerland meant primarily to the Swiss National Science Foundation. The directorate expected the foundation to handle its tasks responsibly.[39] With regard to the federal government's role, the magic words were «overarching areas»: the cantonal directorate committed the state to issues that affected all of Swiss society.

As a mechanism, the special measure was time limited, adapted to socioeconomic dynamics, and customized to dovetail with other mechanisms of funding universities and research. Yet despite the adaptability and customization, the special measure could not escape an already known problem, which is that federal measures are not neutral but have an effect on the autonomy of cantonal universities. Nor can such effects be contained by restricting federal mechanisms, for example, to construction or promotion of technology. From the perspective of the cantons, the federal government's short-term interference remained a sensitive issue, because funding for personnel and research infrastructure influenced specializations, roles, and teaching.

Taken together, these factors suggest that, like other tools of federal university policy, special measures presented universities with the challenge of meeting the demands of both the federal government and the cantons. They fell outside the normal budget process but apart from that their effect was similar to that of other tools of federalism. Special measures and projects showed that it is impossible to draw strict boundaries between research and teaching and between cantons and the federal state. They both rely on formalized planning. The federal government learned that for all the diversity of higher education policy mechanisms and the formality of planning tools, the

39 The Swiss National Science Foundation was founded in 1952. It funds basic research on behalf of the federal government. The Education Directorate addressed the most popular new mechanism of science policy within universities in the 1970s. In research, in teaching, and in administration, projects experimented with structures and collaborations that broke the bounds of conventional institutions, disciplines, and hierarchies. On the introduction of project-centered approaches in research, see Gugerli et al., Transforming the Future, pp. 250–256.

universities could not be managed uniformly because they pursued different directions in research and teaching.

In a third and final section I wish to show why a network for research and higher education was well within the scope of federal science policy. The idea of a network of university computers was first tested within the *courant normal* of cooperative federalism, but it was unknown to most users at the end of the 1970s. Yet the network developed into a productive thought experiment, and in 1985 the concept of a federal network quickly took shape – in advance of the special measures.

To switch or not to switch: «for most of them the idea of a network alone is new»

Much speaks for starting the story of the Swiss network for research and education in the 1970s with a task that was defined by the University Conference's Computing Commission (Commission pour l'informatique). The Computing Commission wanted to find out if it was possible to save money by having the universities' computing centers share their computers' calculating power. Automation and frugality were twin concepts in the federal administration.[40] The assignment was thus in line with the University Conference's mission to coordinate, to report, and to save money. The assignment furthermore accorded with the federal state's procedures for acquiring computing technology. By the end of the 1970s, a diverse landscape of computer use had emerged at Swiss universities. Technically, it was possible to geographically connect distant centers by remote data transmission. Was there also a need for remote data processing? Perhaps even more synergistic effects could be expected. That was the gist of the assignment.[41]

40 See Nick Schwery's contribution in this issue. See also Swiss Federal Archives BAR E7001C#1989/204#145*, Federal Central Office for Organizational Issues, Report on the Use of Electronic Data Processing in the Federal Administration, including an organizational chart with the current status.

41 Jürgen Harms, the first president of the future foundation that supported the network from 1986 onward, referred to this assignment in an interview with the author in March 2019. The final report dealing with the completed assignment can be found top-

It makes sense to start the network's history here, owing to continuities in personnel: some of the computer scientists involved later worked on the Swiss network for higher education and research. Moreover, the objects of desire were identical: the goal was to create a network for computers. Setbacks occurred during this time that had an impact on the network's development. Finally, and most importantly, locating the beginning of the network within the contemporary realities of university policy opens up perspectives on how actors negotiated concepts and practices of federalism in computer-based science and teaching.

In 1978 the Computing Commission created a «networks» working group that set out to conduct interviews with managers and users of university computer centers. The group also sketched out a network prototype. The members learned what the «typical mode of operation of the data centers» looked like. And they wanted to find out how a future network – a «Rechnerverbund-Netzwerk» – could be configured and operated.[42]

The authors of the final report laid out the results of their survey in August 1979. The report shows that the authors were particularly impressed by the diversity of the data centers studied. First, they found different types that mattered from a perspective of sharing computing power. The Federal Institutes in Zurich and Lausanne as well as the University of Zurich were «producers»: they had computing capacities that were accessible to third-party users. Then there were «consumers», computer centers that depended on others to provide calculating power for them. Finally, there was a «self-sufficient» center (Bern) and the «special case» (Basel), which bought computation time from a nonuniversity center.[43]

most in an archival box Albert Kündig handed over to the ETH Zurich University Archives: ETHZ Archives, Akz 2017–19, 7, Swiss University Conference, Final Report of the Working Group Networks, 15 August 1979.

42 ETHZ Archives, Final Report, p. 2. Authors of the report: Bonzon (University of Lausanne), Christen (University of Basel), Erard (University of Neuchâtel), Harms (University of Geneva), Mresse and Mumprecht (University of Zurich), Lutz (Swiss PTT, Bern), Guex (EPFL), Schicker (ETH Zurich), Wyss (Swiss University Conference), Bauknecht (University of Zurich).

43 Ibid., p. 3.

Almost all users – and this came as a big surprise to the authors – were happy with this pragmatic system of (non-)exchanges. At least, they did not wish to connect to another center via a network. According to the report:

> Experienced users of data centers were asked about their ideas on the topic of a computer network. The results showed that for most of them, the idea of a network alone is new. Spontaneously expressed concerns affected the technical and organizational difficulties of user support for computer use (operating system) and program packages in general.[44]

Computer centers were well-established institutions with proper cultures. They informed users about computer programs and documented and collected catalogues so that users could make their choice. A center's personnel consulted users in person.

The knowledge of how to use computing services, once acquired with great effort, leads to a pronounced ‹sedentariness› on the part of the user. This is understandable in view of the large architectural differences between systems of different manufacturers and types. Already the prevailing confusion of language in the manufacturer-specific term categories of electronic data processing favors the isolation of one data center community from others.[45]

The main conclusion the working group drew was that networks weren't suited to meet economic expectations.[46] It was common practice for universities to buy computing power from other academic, administrative, or private computing centers. The users' and managers' concerns pointed to essential features that demanded an immense programming effort if personal contact and consultation were to be substituted by telecommunication and

44 Ibid.

45 Ibid.

46 This conclusion thus reflected the assignment to sketch an economic solution: «Under the main aspect of economic efficiency, possibilities are to be discussed which allow each computer center user to be offered a comprehensive range of computing power and a well-rounded software range under optimal conditions. In addition to a qualitative and quantitative improvement of the IT service, an economically more favorable production of the same is also to be strived for, in that special components can be provided centrally.» ETHZ Archives, Final Report, p. 1.

automation in a multifaceted computer setup. The report predicted that the transformation of a local into a generalized «administration» would be one of the bottlenecks of computer networks: «The contentious issue of user identification, authorizations, and billing, which today is solved individually by each data center, must be replaced by a uniform system throughout Switzerland.»[47]

Peter Schicker, a mathematician who specialized in protocol design and network architecture, reported on behalf of the group to the University Conference's Electronic Data Processing (EDP) Commission.[48] «A general network cannot be considered economical at the moment, but this is not an ‹eternal› decision.» The minutes of the meeting show that Kurt Bauknecht, chairman of the EDP Commission, thanked Schicker and wished to proceed without further ado: «Thanks for the work. The final report will be prepared and transferred to the client by the overall commission.»[49]

The EDP Commission assembled delegates from Swiss universities on a regular basis to share ideas and to report on individual plans for organizing and acquiring technology for data processing. Members of the commission were computer scientists and computer center managers from the universities of Basel, Bern, Fribourg, Geneva, Lausanne, St. Gall, and Zurich as well as from the two federal institutes of technology in Lausanne and Zurich. Bauknecht himself was from the University of Zurich. Kurt Steiner and Urs Hochstrasser, representing the federal Administration, also sat on the committee. Hochstrasser was head of the Federal Office for Education and Science. Steiner represented the Federal Central Office for Organizational Issues and was both experienced in coordinating automation within the federal administration and was responsible for the efficient use of federal funds in any acquisition or operation of IT funded by the Confederation. Representatives

47 Ibid., p. 5.

48 For a contemporary publication by Schicker, see Peter Schicker, The Computer Based Mail Environment – An Overview, in: Computer Networks, 1981, 5: 435–443.

49 Swiss Federal Archives (BAR), BAR E6502-01#1993/126#392*, Minutes of the Meeting of the SHK-EDP Commission, 3 May, Bern 1979. The written version of the report was published in August 1979.

of the University Conference and the Swiss National Science Foundation sat on the board too.[50]

Hochstrasser immediately used the opportunity to learn more: «Is the statement that an entire network is not economical based on expensive transmission or on other factors?» Why was this so? Why did the costs not decrease when material and «computing time» were shared? This question addressed the cost of telecommunication in Switzerland, which was considered high. The question also addressed the area of responsibility of the Federal Central Office for Organizational Issues. It was the office's firm conviction that a correct – i.e., coordinated – use of IT resources would result in added value for all those involved. If the network working group came to a different conclusion with the help of expertise, Hochstrasser would have to listen closely.[51]

But the cost of transferring the data was not the decisive factor. Schicker strove for clarity. He concluded that the decisive factor was heterogeneity, in other words, «the universities' different hardware». The different hardware in turn was due to a plurality of computing technologies. For Schicker, the only way out of the dilemma was sophisticated programming. Alfred Schai, data center manager at ETH Zurich, reiterated that the cost «was due primarily to the complexity of the network», i.e., the mix of hardware and software. Costs that were linked to telecommunications would arise in every network. He suggested that «the final report of the group could serve as a first approach for a future decision».[52]

The report untied a knot. Because the usefulness of sharing hardware and software was fundamentally questioned, cost issues were a distraction. Now a highly productive thought experiment ensued. If costs could not be saved, it did not matter what the PTT charged for data transmission. If networks were ever to become relevant again, costs had to be treated as neither a

50 BAR, SHK-EDP Commission. The commission was an all-male collective of data center managers, professors with educational backgrounds in engineering, natural and economic sciences, and high-ranking administrative personnel.

51 On Steiner and the Federal Central Office for Organizational Issues, see Nick Schwery, Die Maschine regieren. Computer und eidgenössische Bundesverwaltung, 1958–1963, Preprints zur Kulturgeschichte der Technik, ETH Zurich 2018.

52 BAR, SHK-EDP Commission, p. 5.

limiting nor an enabling factor. Networks would have to be viewed as abstract entities for coordinating heterogeneous components. Thus, at least two state institutions were dismissed from the network planners' negotiating table: the Federal Central Office for Organizational Issues, with its savings regime, and the PTT, with its physical infrastructure and culture of standardization. At the same time, the experts interviewed from the computing centers were outvoted: local heterogeneity was not an obstacle, but in fact the goal. It is worth noting that the group had no issue with any specific software or transmission standard at this point, nor the cost of the PTT's so-called leased lines. For them there was a more general lesson to be learned. Networks could open up new possibilities. It was just that they were not made to expand computing functions and to reduce costs at the same time.

Only a few years later, members of Swiss universities had gained network experience, and it almost seemed as though a national network was superfluous. Local heterogeneity was now compatible with global standards. The European Academic Research Net (EARN) enabled Swiss scientists to link up with other networks, computers, and colleagues all over the world. The problem with EARN was that it was supported by IBM, which had pledged its backing only for an initial phase. Consequently, in 1984 a new «network» working group was established, led by Jürgen Harms, a professor at the University of Geneva who had been part of the 1979 group. Harms and his colleagues set about reworking a concept for a «réseau informatique suisse pour les universités et la recherche» – a «Swiss IT network for universities and research». The network was intended to replace EARN in Switzerland in 1988. This network represented coordination on a global, European, federal, and intercantonal level. In April 1985, Rolf Deppeler, secretary general of the Swiss University Conference, was pleased to announce to the Computing Commission that the University Conference had a «great interest in these network issues, and the cantons are willing to finance certain planning work; the federal government will also finance its share».[53]

The network had already found its future users, and the concept enjoyed support among the Computing Commission as well as cantonal and federal

53 BAR E6502-01#1993/126#400*, Minutes of the Joint Meeting of the Commissions for Computer Science and University Libraries, 14 March, Bern 1985, p. 14 f.

donors. The concept was well advanced when the federal government decided to fund a package of special measures. On the same day, Harms and his colleagues from the Computing Commission laid out principles that would apply for the future Swiss network for higher education and research. Harms told his colleagues to be «aware that the establishment of a future network requires a joint effort by the universities». Bauknecht emphasized that «as many building blocks as possible should be adopted» from the Deutsches Forschungsnetzwerk (DFN), the West German equivalent of the Swiss network. The DFN net wasn't yet finished, but the Swiss had a high opinion of the work being carried out by their neighbor. Schai hoped that the Swiss project would align with international standards. «The Confederation is interested in the Computing Commission taking on this role», declared Hochstrasser. «A longer-term concept needs to be worked out to ensure compatibility. When setting up a Swiss network, one must aim at realizing international standards.» Local heterogeneity remained intact; it was compatible with international standards and the federal intent to steer development in the field of computing. The first sketch of the concept was intelligent, and the members of the commission explicitly saw it that way. By the close of the meeting Jürgen Harms and his group were officially in charge of planning the network.[54]

At the same time, initial preparatory work was in process on the special measures. In August 1985, the Swiss University Conference commissioned the Computing Commission to submit proposals in the «immediate future». The conference made clear that there was no time to lose: «Because the concretization and quantification of the measures to be included in this ‹special package› is particularly urgent, CICUS has been commissioned to submit corresponding proposals in the short term.»[55] The commission had formed a

54 BAR E6502-01#1993/126#400*, Minutes of the Meeting of the Informatics Commission (CICUS), 14 March, Bern 1985, pp. 3–6.

55 BAR E6502-01#1993/126#395*, Financial Measures of the Confederation in Favor of the Cantonal Universities within the Framework of the «Special Package Computer Sciences». Proposals of the «Special Measures» Working Group of CICUS, 7/8/1985, p. 1. The paper names several institutions that were already involved in the process or were expected to join soon: the Swiss university conference's plenary assembly, its secretary, its

«special measures» working group that had already received an internal federal discussion paper that listed two «national» components of the package: a national supercomputer, and the same network for universities and research that was debated by the commission. The University Conference's working group «special measures» accepted both and wrote in the amount of 15 million Swiss francs to be paid by the federal government for the network, noting that it preferred joint sponsorship of cantons and federal government. The working group placed particular emphasis on the cantonal universities: «Care must be taken to ensure that the needs of the cantonal universities are taken into account with regard to the use of the national supercomputer.» It was important «that planning be done in close cooperation between the federal government and the ETH Board on the one hand and the [Swiss University Conference] and [its Computing Commission] on the other.»[56] In mid-1985, the working group proposed to allocate funds for temporary hiring of staff in computer science major and minor programs for both teaching and research. This measure could be clearly distinguished from other potential funding opportunities. Moreover, the amount to be allocated for it was «reasonable. [...] Staff positions and acquisitions for computing centers and computing services, as well as workstations and computer acquisitions for departments outside of computing» would thus clearly not be eligible. The University Conference wrote in 20 million Swiss francs for personnel at cantonal universities and another 12 million francs for the acquisition of workstations.[57]

Conclusion: cooperation in practice

Once the special measures were launched, the network was quickly assigned a special position within them: the promise of a future network of scientific computers transformed geographically, socially, and technically disparate projects into a consolidated «national» package. From a historical perspec-

Computing Commission and its «special measures» working group; the Federal Office for Education and Science; the Swiss National Science Foundation; ETH Board.

56 BAR, «Special Measures», p. 2.

57 BAR, «Special Measures», pp. 7 and 8.

tive, the special measures are an example of the possibilities of federal states to bundle together something that is already there and thereby achieve synergistic effects. The Swiss special measures make sense when seen in the historical context of Swiss investment in economic change since the 1970s. The measures have fully exploited the coordinative potential of a mechanism known as the «network». They offered a means of coordinating a «post-industrial» society that had not yet taken shape, especially at the supra-regional level. The network mechanism thus soon took on a dual nature: it was ideal and material at the same time.[58]

The special measures did not alter the state's role as funder. They reinforced the view that the federal government was in charge of «overarching» issues. By the same token, the individual measures in the package did promote geographical expansion of a digital federalism in *statu nascendi.* When the decision was finally taken in 1989 to establish a national center for supercomputing in the hitherto academically untapped canton of Ticino, it was a pixelated graphical representation of a nationwide digital network that made this step appear viable and visionary. Telecommunications networking even suggested that measures targeting the national level had the effect of stimulating international relations.

Yet the course of history that led to this sort of integration of the package of special measures was unpredictable. It was not clear at all to contemporary computer scientists in the late 1970s what a national computer network should look like. The successful match of computer science and federalism was possible only after the idea of an efficient federal-cantonal computer network had been discarded and replaced by the stable expectation of synergies in scientific communication. Recapitulating the development that led to the special measures reveals a few more surprises. It is important to appreciate these surprises, because they confirm that the concept of cooperative federalism was successfully rehabilitated and open to international exchange on many scales.

---

58 This finding of the imaginative power of networks corresponds to results of international research on national networks. See, for example, Eden Medina, Cybernetic Revolutionaries: Technology and Politics in Allende's Chile, Cambridge MA 2014. Paolo Bory, The Internet Myth: From the Internet Imaginary to Network Ideologies, London 2020.

Within the web of national relationships established by the cantons and the federal state, the actors and institutions involved moved smoothly to unexpected positions by mutual agreement. The planning of the network was initiated by the University Conference, an institution that represented both the federal state and the cantons. Nevertheless, the federal government put it on the list of «its» special measures, and was willing to pay a bigger share and to play a major role in developing the network. Furthermore, networking was not seen as a matter of physical connections between computers and the PTT.

The involvement of federal authority changed considerably while the network was taking shape: «Bern's» influence changed because the national and federal authority of the PTT and the Central Offices for Automation declined. At the same time, the future outcome of an investment in universities wasn't original but rather was intended to copy an international, perhaps German model. Another surprise was expressed by Jürgen Harms, the first president of the future SWITCH foundation that was to establish and maintain the network of the same name. He stressed how much effort was required to create this organization – a foundation that was carried by cantons and the federal government – «that could purposefully support and promote cooperation between universities». As far as the organization was concerned, «role models from the international arena were of little help. [...] Practically every country had its own boundary conditions [...] and came up with its own solutions».[59] Harms was later actively involved in enlarging the geographical, social, and political space of higher education by offering computer access to users at universities of applied sciences and by guaranteeing heterogeneity to operators of academic computing technology.

The future SWITCH computer network and its supporting organization, the SWITCH Foundation, were innovative, and hence constituted special measures relating to university policy that relied on institutions and helped to establish rules and boards for discussion and decision-making. Thinking and shaping cooperation is an important task for higher education and research institutions. This implies that planning and implementing a network was political and that it was, as Harms has put it, located on «this

59 Author's interview with Jürgen Harms in Geneva, 26/3/2019.

side» of the state and legal structures. With regard to the intended effects of the special measures – education, training, and managing structural change – the network most likely performed poorly or at least in a way that cannot be measured. Moreover, the emphasis of the special measures on temporary employment certainly did not alleviate the universities' personnel problem, especially when seen from the perspective of the employees. The positive societal impact is to be found somewhere else. The computer sciences today are still most effective when they invest time and resources in elucidating society, technology, and organizations. In this sense, computing is social precisely when it turns its gaze on itself. Computer science, like any science, has the potential to be unelitist and excellent at the same time.

Supercomputing and the emergence of digital federalism

David Gugerli, Ricky Wichum

Abstract

Supercomputing has been an integral feature of science policy and programs since the late 20th century. As such, it has extended the capacity for calculation and enhanced the power of simulation. The enormous amount of money spent on high-performance computers implies that these machines are important because they are expensive – the priority given to them stimulates both conspicuous computing and demonstrative government spending. Providing a specialized scientific community with cutting-edge computational power benefits from centralized resources. Consequently, it seems natural that allocation of these budgets be left to the highest levels of government decision-making. Yet, perhaps surprisingly, this does not mean that federal political systems with many decision levels are automatically handicapped vis-à-vis the development of supercomputing. Our study focusing on the southern part of Germany shows that, historically, supercomputing also developed as an essential cornerstone of digital federalism, an emergent field of technoscientific practice.

Introduction

Supercomputers are the fastest and most powerful computing systems in the world. They offer much higher speed and performance than any other class of computers. Supercomputing also means the allocation of an extreme amount of resources in a single and very complex computing center. This leads to operating costs no market could possibly absorb. Hence, supercomputing has always been a playground for powerful, splendid governments and their technoscientific programs. Like military might, supercomputing came under the national purview and was the privileged realm of strong governments whose centralistic allocation of financial, human, and technological resources produced representational effects. Supercomputing was organized as a worldwide competition for processing power and a very earnest game of «conspicuous computing».[1]

Three decades ago, Donald MacKenzie asserted that the Los Alamos and Lawrence Livermore National Laboratories exclusively sponsored and shaped the development of supercomputing architecture. It was the labs' nuclear programs that set computing standards: they defined the main criterion for deployable computing capacity as the number of floating point operations per second (FLOPS).[2] In the 1980s, FLOPS became an early means of comparing the computing power of any calculating sovereign anywhere in the world. Only later, in 1993, did the Top500 list offer a slightly more sophisticated benchmark for state-of-the-art supercomputing by ranking all centers. Over the last quarter of a century, the Top500 list has closely tracked the field of high-performance computing (HPC) and reported where «computing at the limit of computability» is happening.[3]

Despite its association with politics, power, and sovereignty, almost no attention has been paid to supercomputing by science and technology studies. Whereas most scholars in the field did discuss the centralizing and decentralizing effects of computers in general, they typically claimed that «the» computer and «the» network dissolved the geographic, political, and legal boundaries of the nation-state. Moreover, they argued that this dissolution led to the formation of new global infrastructures and new spaces of a fragmented and heterogeneous, yet very powerful political authority.[4] However, any increase in autonomy at the periphery might simply have been illusory. According to conventional science and technology studies, hardware and software remain a source of central authority even at its remotest ends, since

1 Computing, not consumption was the demonstrative purpose of this game.

2 Donald MacKenzie, The Influence of the Los Alamos and Livermore National Laboratories on the Development of Supercomputing, in: IEEE Annnals of the History of Computing, 1991, 13(2): 179–201.

3 David Gugerli, Supercomputer – an der Grenze der Berechenbarkeit, in: Merkur. Deutsche Zeitschrift für europäisches Denken, 2019, 73(846): 53–59.

4 Jean-François Lyotard, La condition postmoderne. Rapport sur le savoir, Paris 1979; Saskia Sassen, Losing Control: Sovereignty in the Age of Globalization, University Seminars: Leonard Hastings Schoff Memorial Lectures, New York 2015; Keller Easterling, Extrastatecraft, London 2014.

they provide the sovereign with the silent language of digital codes, which are difficult to contradict.[5] Consequently, science and technology studies encouraged analysis of digital and mathematical games of language in the development of algorithms and simulations; they highlighted the importance of a political economy of hardware and software providers; and they considered the importance of computing centers as the focal points of data-driven knowledge production.[6] Obviously, there was no incentive to distinguish between supercomputers and computers in general. Insights into the (de)centralizing effects of the computer were necessarily as true for standard computing as for computing in extremis, i.e., for HPC.

In what follows, we argue that fully understanding the politics, power, and sovereignty of supercomputing requires introducing more distinctions to allow a more subtle analysis. First, we wish to reintroduce the distinction between «the computer» and «the supercomputer». Second, we wish to differentiate the technopolitical notions of center and periphery by discussing the configuration of three historical forms of digital federalism.[7] In order to demonstrate this empirically, we draw on our case study on the history of technology of supercomputing at the University of Stuttgart.[8] In this study, we very much had the impression that the history of HPC must take into consideration a wide array of stakeholders acting in what we call «digital fed-

5 Friedrich Kittler, Protected Mode, in: Norbert Bolz et al. (eds.), Computer als Medium, Munich 1994, pp. 209–220.

6 Lucas Introna, David Wood, Picturing Algorithmic Surveillance: The Politics of Facial Recognition Systems, in: Surveillance & Society, 2002, 2(2/3): 177–198; Tung-Hui Hu, A Prehistory of the Cloud, Cambridge 2015; Malte Ziewitz, Governing Algorithms: Myth, Mess, and Methods, in: Science, Technology, & Human Values, 2015, 41 (1): 3–16; Robert Seyfert, Jonathan Roberge (eds.), Algorithmuskulturen. Über die rechnerische Konstruktion der Wirklichkeit, Bielefeld 2017; Sven Opitz, Simulating the World: The Digital Enactment of Pandemics as a Mode of Global Self-Observation, in: European Journal of Social Theory, 2017, 20 (3): 392–416.

7 However, we will not distinguish between *supercomputing* and the more operational term *high-performance computing*, since both terms are similarly different from other forms of computing, e.g., computing supported by mainframe computers, data processing computers, file servers, or computers used exclusively for network maintenance.

8 David Gugerli, Ricky Wichum, Simulation for All: The Politics of Supercomputing in Stuttgart, Zurich 2021.

eralism». Therefore, we focus on the distribution of authority among the many different layers of a federal political system and the many and highly differentiated public spheres.[9]

We will argue that HPC is not so much the effect of a centralistic government's decision gracefully offered to its scientific, military, and industrial subjects, but rather the result of permanent and complex negotiations in a dynamic, multileveled political context of technoscience.[10] We distinguish three successively dominant forms of digital federalism in Germany in the late 20th century. The first was a program oriented at federal distribution (mainly in the 1970s). The second – in evidence roughly from the mid-1980s to the mid-1990s – was a form of digital federalism that sought to allow supercomputing the power to grow from regional competitiveness. Finally, in the second half of the 1990s, a new digital federalism started to emerge. It materialized as a product of various special «trading zones». These trading zones primarily centered on organization, access to supercomputing, and even evaluation of architectural options. Some of them persisted only temporarily; others grew in importance and became quite stable. In terms of their institutional framing, most trading zones varied from one working group to the next, involving a variety of sponsors and many organizations, thus maintaining a growing network of both federal and regional actors. All produced traces of their deliberations – vast materials we were able to access in the Stuttgart University Archives, and the personal document collections of former and current employees at the High Performance Computing Center Stuttgart.

Distributing funds under digital federalism

In the 1970s, computing became a matter of course in German universities and research institutions. This computational normalcy was largely the result of a federal policy aimed at overcoming American supremacy in software

9 For a similar approach in the political sciences, see Jenna Bednar, The Political Science of Federalism, in: Annual Review of Law and Social Science, 2011, 7: 269–288.

10 Peter Galison, Trading Zone: Coordinating Action and Belief, in: Mario Biagioli (ed.), The Science Studies Reader, London 1999, pp. 137–160.

competence, equipment, and computing power.[11] «Germany» should catch up with «the US» (or «Siemens» with «IBM») by building up a strong computational infrastructure and genuine computing competence. The federal policy for technoscience operated through an established distribution network and intensive cooperation between the federal state and its member states.[12] Federal programs in technoscience were consensus oriented and essentially based on the principle of uniform funding. Two agencies of the federal state, the Ministry of Research and Technology and the German Research Foundation (DFG), defined the terms for allocating subsidies.[13] They carefully orchestrated interactions between the federal government and the member states. There were two different channels of technoscientific development: the University Construction Act (Hochschulbauförderungsgesetz, from 1972), which governed construction of new university buildings, and a trio of funding programs for «data processing» (1973–1981).[14]

These programs did not distinguish between different machines. For the programs as such, it was irrelevant whether the funds they provided went to support a mainframe computer, a programmable data processor, or a high-performance machine. The task was to provide science and research with the computing capacity they arguably needed. Institutions got the machine and

11 Hartmut Petzold, Moderne Rechenkünstler. Die Industrialisierung der Rechentechnik in Deutschland, Munich 1992; Johannes Bähr, Die «amerikanische Herausforderung». Anfänge der Technologiepolitik in der Bundesrepublik Deutschland, in: Archiv für Sozialgeschichte, 1995, 35: 115–130.

12 Fritz Wilhelm Scharpf et al., Politikverflechtung: Theorie und Empirie des kooperativen Föderalismus in der Bundesrepublik, Kronberg/Ts. 1976; Andreas Stucke, Staatliche Akteure in der Wissenschaftspolitik, in: Dagmar Simon et al. (eds.), Handbuch Wissenschaftspolitik, Wiesbaden 2016.

13 Ulf Hashagen, Computers for Science – Scientific Computing and Computer Science in the German Scientific System 1870–1970, in: Mark Walker et al. (eds.), The German Research Foundation 1920–1970: Funding Poised between Science and Politics, Stuttgart 2013, pp. 135–150.

14 Timo Leimbach, Die Softwarebranche in Deutschland. Entwicklung eines Innovationssystems zwischen Forschung, Markt, Anwendung und Politik von 1950 bis heute, Stuttgart 2011, p. 166; Bernd Reuse, Roland Vollmar (eds.), Informatikforschung in Deutschland, Berlin 2008.

the computing power whose acquisition they could justify in a funding application.[15]

The federal technoscientific programs certainly had an impact. The hardware used at German universities became faster and the available software more differentiated, while the number of users from different disciplines skyrocketed. The availability of additional computing capacity even led to new study programs for computer science and for applied computer science, in Stuttgart and elsewhere.[16] However, toward the end of the 1970s federal programs began to lose some of their coordinating power. It was difficult to maintain a coherent set of rules for future development. The programs and mechanisms for distributing funds ran into hurdles.[17]

In 1977, as previously planned, the University of Stuttgart applied for a new, powerful mainframe computer – a Cyber 175 – to be financed by the DFG.[18] The proposal failed miserably to convince the reviewers. While the importance of computing per se was beyond any doubt, there was no consensus among these experts on how a normal university was to keep up with increasing computational needs. The DFG did not believe in Stuttgart's long-term expansion course, nor did they buy the argument of a growing demand for centralized computing power.[19] Moreover, there were serious doubts

15 Joachim Mönkediek, Rückblick auf die Förderung der Informationstechnologie an Hochschulen durch Bund und Länder, in: Hannes Hartenstein (ed.), Informationstechnologie und ihr Management im Wissenschaftsbereich. Festschrift für Prof. Dr. Wilfried Juling, Karlsruhe 2009, pp. 23–28.

16 Leimbach, Softwarebranche, pp. 218–219; Christine Pieper, Das «Überregionale Forschungsprogramm Informatik» (ÜRF). Ein Beitrag zur Etablierung des Studienfachs Informatik an den Hochschulen der Bundesrepublik Deutschland (1970er und 1980er Jahre), in: Technikgeschichte, 2008, 75(1): 3–31; Bernd Reuse, Roland Vollmar (eds.), Informatikforschung in Deutschland, Berlin 2008. On the early connection between computer-based research and teaching, see Christoph Hoffmann, Eine Maschine und ihr Betrieb: Zur Gründung des Recheninstituts der Technischen Hochschule Stuttgart, 1956–1964, in: Barbara Büscher et al. (eds.), Ästhetik als Programm: Max Bense/Daten und Streuungen, Berlin 2004, pp. 118–129.

17 Leimbach, Softwarebranche, pp. 182 ff.

18 Stuttgart University Archives (UASt), Universität Stuttgart, Rechenschaftsbericht des Rektors 1975/76, p. 246; Ibid. 1976/77, p. 242.

19 Ibid. 1977/1978, p. 303.

about the preferred hardware provider.[20] Thus, it became difficult to defend the well-established mainframe solution for a data processing factory at the center and a growing number of terminal stations at the periphery. The existing installation of a CD 6600 coupled with a Cyber 174 certainly integrated many different applications. However, there was nothing spectacular about the infrastructure and the planned applications; there were no technological thrills and barely any scientific challenges.

The DFG simply declined to fund the planned expansion of the computing center at the University of Stuttgart. Instead, it started to support programmable data processors and other smaller machines in the university's most active research institutes.[21] Small machines became so beautiful that one could no longer successfully argue with the economies of scale of centrally supplied computing power.[22] Even such strong defenders of computing centers as the editors of the journal *Rechenzentrum* publicly mentioned the imminent possibility of a sudden «death of the computing center». In the context of the midi-computer hype and the appeal of distributed computing, directors of existing computing centers had to face the vaporization of their budgets. A fair amount of conceptual homework must have featured on their to-do-lists.[23]

If the computing center was still to play a key role in the future development of academic computing, it would have to be emancipated from its data factory model. In order to not succumb to a distributed processing scheme, the center had to defend its centrality with the help of a new, but genuinely

20 See UASt, Deutsche Forschungsgemeinschaft, 12. Mai 1978, Stellungnahme der Kommission für Rechenanlagen zur Anfrage des Landes Baden-Württemberg vom 26. Juli 1977 bezüglich Erweiterung des Regionalen Rechenzentrums Stuttgart im Programm zur Errichtung Regionaler Rechenzentren (Az.: 375224/2/77), 12 May 1978.

21 Karl-Gottfried Reinsch, Strukturveränderungen des Rechenzentrums und seiner Benutzer, in: Das Rechenzentrum, 1982, 31: 171–175, on p. 171.

22 Karl-Gottfried Reinsch, Regionale Großrechenzentren zur Versorgung japanischer Universitäten. Praxis der Informationsverarbeitung und Kommunikation, in: Das Rechenzentrum, 1980, 3(2): 73–79, on p. 79.

23 Cf. Das Rechenzentrum, 1978, 2, pp. 59 and 60.

central service. HPC was definitely one such possibility that many scientists could not ignore.[24]

The idea was simple, but the implementation a wearisome endeavor. It also entailed learning the hard way, since supercomputing implied reshaping the hitherto well-established relations between all stakeholders – on campus, in industry, and in the federal and local governments. Buying a supercomputer could mean all sorts of things. Some people at the center even floated the extravagant idea of buying a supercomputer and eventually using it as a substitute for the old CD 6600 mainframe.[25]

The most difficult problem for the University of Stuttgart was posed by its regional competitor, the Technical University of Karlsruhe, after the DFG flatly rejected proposals from both universities in 1980. A supercomputer is a very expensive, very exclusive tool that was at the limit of the agency's budget. The DFG played hardball, refusing to decide which university in Baden-Württemberg should get a supercomputer.[26] Instead, the agency stipulated that the universities would have to develop a means of coordinating and distributing their services. By restructuring its approach to distributing funds, it sought to obtain a long-term strategic advantage over its clients. If Karlsruhe or Stuttgart was the question, the choice between them was not a matter of concern. But the question was an important one for the state of Baden-Württemberg, which had to choose an institution within the state; they could not blame the federal government for a wrong decision. All the while, each of the universities was going to great lengths to justify its burning need for a supercomputer. That is how the DFG succeeded in developing new spaces of negotiation for distributing federal subsidies and found a way to insert these trading zones into the technoscientific landscape of the member state in question. The qualitative improvement in terms of legitimacy of federal

24 See UASt, Reinsch, Strukturveränderungen. Universität Stuttgart, Wissenschaftlicher Grösstrechner für das Land Baden-Württemberg. Antrag zur Realisierung am Regionalen Rechenzentrum der Universität Stuttgart, Stuttgart 15/3/1980.

25 UASt, Rechenschaftsbericht des Rektors 1978–1980, p. 171.

26 Peter Sandner, ALWR-BW. Arbeitskreis der Leiter der Universitätsrechenzentren in Baden-Württemberg, in: PIK, 2008, 31(3): 193–198, on p. 193.

sponsorship was considerable. Moreover, it strengthened the distinction that supercomputing brought to the winning team.[27]

Negotiations are no substitute for decisions. Stuttgart and Karlsruhe were surely each bound to lose their wager if they failed to reduce the estimated costs for their future supercomputers. Obviously, neither the University of Karlsruhe nor the University of Stuttgart could be arm-twisted into backing down. Instead, both universities reduced the federal part of their budget by half and sought to compensate the remaining half through contracts with future external users. For the subsidizing DFG, this was a win-win-win situation. It gained two supercomputers at two different places for the price of one, doubling its subsidizing effect and enhancing the legitimacy of its policy. Science policy had positive effects not only for public universities but also for industrial enterprises. The industrial partners finally began to invest in supercomputing, the universities guaranteed its operation, and the computing centers maintained all necessary network connections.[28]

The simultaneous acquisition of a supercomputer in Karlsruhe and in Stuttgart produced a trading zone in Baden-Württemberg that achieved consensus on seemingly contradictory or mutually excluding interests. First, the new digital federalism created a new arena of decision-making on a regional level. The university rectors and the regional ministry of science and arts, together with the heads of the computing centers, negotiated the procurement and organization of supercomputing in Baden-Württemberg. Second, digital federalism increased the autonomy of both the state and the federal government. While the federal government placed supercomputing within its distribution program and thus legitimized and consolidated the program as a whole, the federal state and its universities also gained new formative power. By no means did these developments herald a spectacular ushering in of a

27 The rectors' conference (Landesrektorenkonferenz) in Baden-Württemberg had to establish a working group to mediate the conflict over scarce resources, keep it at a controllable level, and work out an acceptable consensus that could then be submitted again «to the federal government». Ibid. The working group was the nucleus for the «Arbeitskreis der Leiter der Universitätsrechenzentren in Baden-Württemberg», founded in 1981, whose members concluded the «Peace Treaty of the Lake of Constance» at their first meeting.

28 Ibid.

new supercomputing era in Baden-Württemberg. However, the University of Stuttgart had managed – at least formally – to join the club, to handle the new negotiation culture imposed by the federal funding agency, and to maintain the centrality of its computing center with the help of a relatively cheap Cray-1.[29]

Regional competition in digital federalism

In fall 1983, the new Stuttgart supercomputer found its place of operation in the former main kitchen on the Vaihingen campus, directly wired to the existing computing center. Although the Cray-1 no longer represented the cutting edge of supercomputing, Stuttgart's formal participation in the supercomputing game greatly altered perspectives and expectations. The university's computing center forgot the good old times of mainframe computing. Even more important was finding out that the scientific and economic coordination provided by the federal funding agencies was probably not the sole source of orientation for local strategies and developments. The complex acquisition process of the Cray-1 made clear the importance of a whole array of alternative allies. Consequently, in the mid-1980s, the University of Stuttgart developed a close cooperation with the state of Baden-Württemberg's government and its restless minister president, Lothar Späth. The university discovered the benefits of setting an independent agenda for its future development in supercomputing.[30] The field was about to play an important political role as an argumentative resource in the context of an increasingly dynamic and competitive German federalism.[31] Späth announced

29 The Cray-1 made computer history when it was launched in 1976 with its unprecedented computing capacity. The first computing center in Germany to operate a Cray-1 was the Max Planck Institute for Plasma Physics in Garching (1979). Friedel Hoßfeld, Vektorrechner, in: Phys. Bl, 1984, 40(8): 280–281.

30 Franz Effenberger, Lothar Späths Forschungsförderung und Technologiepolitik am Beispiel der Universität Stuttgart, Ubstadt-Weiher 2020.

31 In a programmatic statement of the Science Council in 1985: «Wettbewerb setzt zunächst ein gewisses Maß an Handlungsfreiheit für die am Wettbewerb Beteiligten voraus. Wer sich im Wettbewerb bewähren soII, muß das Recht und die Möglichkeit haben,

his «pivot to the future», insisted on the potential for shaping conditions, and strongly advocated a combined science and economic policy.[32]

In June 1985, Späth traveled to the US, where he visited Cray Inc. in Minneapolis. This was an excellent opportunity to demonstrate what the pivot to the future implied. Späth audaciously decided to buy a cutting-edge Cray-2, with no concern for the complicated federal coordination mechanisms that applied in acquiring a supercomputer in Germany. The contract with Cray was a demonstrative act of governmental decision-making. The provider even assured that he would not deliver another machine to any other German university for a year. Competition apparently went hand in hand with new forms of market protection. Europe's first Cray-2 was built out in Stuttgart and was to be exclusively operated by the Stuttgart computing center. However, buying the machine proved much easier than operating it. The old Cray-1 was well established and handled calculations for most of Stuttgart's supercomputing projects. It was not easy to find new clients whose computing needs required access to the Cray-2, and it was simply impossible at the price the University of Stuttgart and the government of Baden-Württemberg settled on to justify their lonely decision. The free-market economy

nach eigener Entscheidung individuelle Leistungen zu erbringen und dabei auf die Signale des Wettbewerbsmechanismus zu reagieren» («Competition presupposes a certain degree of freedom of action for those involved in the competition. To be competitive, they must have the right and the opportunity to decide whether to provide individual services and to respond to the demand for resources»). Wissenschaftsrat, Wettbewerbsempfehlung des WR. Empfehlungen zum Wettbewerb im deutschen Hochschulsystem, Cologne 1985, p. 7.

32 Lothar Späth, Wende in die Zukunft. Die Bundesrepublik auf dem Weg in die Informationsgesellschaft, Spiegel-Buch, Reinbek bei Hamburg 1985. On the reorganization of the rules of science policy in the DFG around the middle of the 1980s, cf. Alexander Gall, «Bundesliga-Spielregeln in der Wissenschaftspolitik». Föderalismus und Forschungspolitik zur Mikroelektronik, in: Johannes Abele et al. (eds.), Innovationskulturen und Fortschrittserwartungen im geteilten Deutschland, Cologne 2001, pp. 147–164. On the discovery of the shapability of society, see Adalbert Evers and Helga Nowotny, Über den Umgang mit Unsicherheit. Die Entdeckung der Gestaltbarkeit von Gesellschaft, Frankfurt am Main 1987.

makes for an exhilarating topic of conversation until the market declines to pay the bill.[33]

For the Stuttgart computing center, the Cray-2 adventure entailed considerable work at the conceptual and communicative level. The old regime of cooperative allocation of federal funds was increasingly being replaced by intensive regional networking and an increased need to publicly explain the role, advantages, and operative principles of supercomputing. This was Roland Rühle's task. The new scientific director of the computing center was keen to present HPC in Stuttgart as the university's central platform for simulation to meet the intellectual and operational needs of the city's research institutes and industrial enterprises. This constituted a broad range of clients with one common denominator: a strong interest in evidence and demonstration. Hence, the coupling of simulation and visualization techniques, which around 1990 represented a novel approach in supercomputing.[34] Visual demonstrations began determining the semantics even of a supercomputer's number-crunching operations. Stuttgart went so far as to claim that its Cray-2 opened «a completely new dimension of technoscientific simulation».[35] Instead of staring at «endless, barely understandable rows and columns of numbers», users were provided with «visualizations of technological processes and scientific models».[36] The pictorial evidence of any kind of supercomputer-based simulations found its way into countless project descriptions, conference slides, and published papers. Images, not tables, produced the insights into an artificial world created by the supercomputer.

Pictorial competence was good news and was welcomed by sponsors and users. Images were a helpful means of reducing cognitive complexity, and they propelled the development of theoretical models in engineering. Wherever interdisciplinary cooperation of specialists was as critical as it was

33 UASt, Karl-Gottfried Reinsch, Bericht Betriebszeitraum CRAY-2 vom November 1986 – Oktober 1987, p. 9.

34 See, for example, Ulrich Lang and Roland Rühle, Visualisierung von Supercomputerberechnungen am netzintegrierten Ingenieursarbeitsplatz, in: Hans-Werner Meuer (ed.), Heterogene Netze und Supercomputer, Berlin 1992, pp. 121–133, on p. 125.

35 UASt, Universität Stuttgart, Antrag der Universität Stuttgart auf Beschaffung eines Höchstleistungsrechnersystems als Nachfolgerechner für die Cray-2 1992, p. 5.

36 Ibid.

in the context of a computing center, calculation, representation, supercomputing, and visualization became amalgamated into a coherent analytical tool.[37]

This methodological shift changed the role of the computing center. It became known for its centralized resources and strict rules of engagement. Moreover, it was run as an ensemble, if not a confederation, of machines, procedures, applications, and analytical tools. The center's personnel learned to expect and to exploit the heterogeneity of its resources. The center was building bridges between these diverse resources and coming to grips with the thematic and disciplinary differences of its users. In fact, the computing center remained a center precisely because it was able to integrate the diversity of its machines, networks, applications, and users. The Cray-2 was a prominent and privileged component in this game. Its privileging effects, however, stemmed from many other machines that protected the supercomputer from tasks other computers could handle just as well.[38] In Stuttgart, the Cray-2 «produced calculations» exclusively. «All other services, such as file servers, backup servers, dialog servers, graphic servers, printer servers, and network servers ran on pre-processing machines».[39] The research institutes connected their midi-computers, workstations, and desktops to the computing center through local area networks (e.g., DECnet). The computing center in turn also coordinated all connections to the European academic re-

37 Ibid., p. 312, and UASt, Universität Stuttgart, Forschungsbericht 1995/96, p. 348.

38 The principle of relieving the load on particularly powerful computers was applied as early as the mid-1960s at NASA's control center in Houston. At Mission Control Center, there were input computers, backup computers, systems for graphic evaluation, and around 2000 employees who reduced the amount of central computing and provided additional analytical loops or electromechanical and optical substitutes for the main computers. Cf. David Gugerli, Wie die Welt in den Computer kam. Zur Entstehung digitaler Wirklichkeit, Frankfurt am Main 2018, pp. 88–105.

39 UASt, Rechenzentrum Universität Stuttgart [Das Rechenzentrum der Universität Stuttgart 1986. Broschüre aus Anlass der bevorstehenden Auslieferung einer Cray-2 im September 1986], Stuttgart 1986.

search network, the German research network, and an experimental ISDN network developed by the federal post office (*Bundespost*).[40]

Increasing user diversity was yet another strategy for the Stuttgart computing center that helped improve the load of its Cray-2. The center managed to incorporate new regional user groups outside the university. One of the most prominent cases was the automotive industry. Porsche's research and development department simulated crash tests for sports cars.[41] Moreover, regional midsize manufacturers became members of Stuttgart's user community.[42] The state of Baden-Württemberg, the city and University of Freiburg, the Chamber of Industry and Commerce Southern Upper Rhine, and the Regio-Gesellschaft in Freiburg, a regional organization, founded High Tech Computerdienste Oberrhein GmbH, a services company.[43]

The increased diversity of services laid bare two principal problems of digital federalism: users (researchers, companies) and the public (government, citizens, and the media). By the end of the 1980s, the center had to find a means of keeping users at secure distance from the machine. The center tried to develop a protected mode of operation and offered users a uniform view on the center's infrastructure. Services were not available until a user accepted centrally defined procedures. At the same time, the center's representatives were well aware that Lothar Späth's trip to Minneapolis

40 A brochure published by the center in August 1987 stresses the diversity of network access: «Die Anlagen des Rechenzentrums sind mit Hilfe verschiedener Technologien (IBM-, DECnet-, TCP/IP-, OSI-Protokolle) über lokale und überregionale Netze (Ethernet, HYPERchannel, DATEX-P/DFN, EARN) erreichbar» («The facilities of the computer center can be accessed using various technologies (IBM, DECnet, TCP/IP, OSI protocols) via local and national networks (Ethernet, HYPERchannel, DATEX-P/DFN, EARN)»)., UASt.

41 Sauer Papers, Stürme aus dem Rechner, in: hightech 5/89, p. 5, HLRS.

42 Sauer Papers, Breisgaumetropole schließt sich Stuttgarter Superrechner an, in: Badisches Tagblatt, 20/9/1988, HLRS.

43 The establishment of this regional distribution company corresponded to the demand of Stuttgart University's chancellor Blum in the expert seminar for new organizational models for distributing the computing capacity of the Cray-1. Cf. UASt, Jürgen Blum, Vorwort, in: Jürgen Blum (ed.), Höchstleistungsrechner. Anwendung. Finanzierung. Organisation, Stuttgart 1985, pp. 2–3.

transformed the well-equipped computing center in Stuttgart into a concern for a wider public. *Computer Zeitung* magazine was a critical observer of the computing center's performance. In the spring of 1987, it published a biting article under the title: «The computing center of the University of Stuttgart acquired a Cray-2. Then the trouble started».[44] It was clear that science policy desperately needed to pay more attention to public relations. Operating a publicly funded supercomputer entailed communicative requirements far beyond the proposal, justification, and reporting scheme of a standard project in the academic world.

Trading zones of a new digital federalism

Much to everyone's surprise, the Cray-1 in Stuttgart remained in operation even after the installation of the Cray-2. When the old machine was finally retired in summer 1987, the new machine still ran with a very modest workload of barely 30 percent. It took half a year to increase its workload to «almost» 70 percent.[45] Nevertheless, there was no serious doubt that acquisition of a next-generation Cray-3 was inevitable. For one thing, the future of supercomputing was promising. The computing community strongly believed that there was a growing and almost unlimited demand for future computing applications in science and industry. Hence, it was still necessary to acquire supercomputers with greater processing power.[46] Baden-Württemberg promptly secured a contract that made sure the first Cray-3 in Europe would

44 Computer Zeitung, 4 March 1987.

45 Cf. UASt, Karl-Gottfried Reinsch, Bericht Betriebszeitraum CRAY-2 vom November 1986 – Oktober 1987, p. 9. This also led to a conflict with the building authorities, as the additional power requirement due to the double operation could only be covered thanks to the provisional installation of a transformer that was needed elsewhere, leading to either uncertainties in the building planning or additional investment costs. UASt, Universitätsbauamt an Rechenzentrum der Universität Stuttgart, 13. Oktober 1986.

46 Hans-Werner Meuer, Parallelrechner – bringen die 90er Jahre den Durchbruch? (erschienen in PIK 1/1990), in: Hans-Werner Meuer (ed.), Supercomputer 1986–1990. Anwendungen, Architekturen, Trends, Munich 1992, pp. 387–393, on p. 388.

come to Stuttgart.[47] Most of the experts in the supercomputing field were convinced that this machine – like its predecessor – would dramatically shift the «limit of computability».[48]

The Cray-3 changed the conditions of digital federalism, even though Cray Inc. ultimately abandoned the project and never built the machine. Supercomputing became a demonstration object for the power of digital federalism beyond its actual materialization. Supercomputers were powerful instruments when operated in computing centers, but also on paper in cutting-edge research projects and in government strategies. The demonstrative effects on paper extended into published records. This was the case in June 1993 when the Mannheim Supercomputer Seminar published the first Top500 ranking. The list gave the supercomputer a paper-based source of demonstrative power. Just as *Forbes* published a list of the 400 wealthiest people in the US, the Top500 aimed at ranking the most powerful computers in the world.[49] This ranking gained influence by ignoring many things. For example, the supercomputer list did not track computer costs, it took no notice of the importance of providers, nor was it keen to report on the special architectonic advantages of a given supercomputer. Henceforth, supercomputers would be exclusively distinguished according to their computing power as determined by the LINPACK Benchmark.[50] If the benchmark defined the power of a machine, then the institutional site, the knowledge of the crew, and the government sponsoring its computing center were of secondary significance. From 1993 onward, it was not the organizational context that defined a supercomputer, but rather the supercomputer that defined its context.

---

47 UASt, Correspondence, John Rollwagen (Cray) an den Ministerpräsidenten des Landes Baden-Württemberg Herrn Lothar Späth, undated (probably 1985).

48 Robert Übelmesser, Die neuen Supercomputer von Cray, in: Hans-Werner Meuer (ed.), Supercomputer '89. Anwendungen, Architekturen, Trends, Berlin 1989, pp. 31–42, on p. 36.

49 See Harald Lux 1997: «Hans-Werner Meuer ist Herr über die Hitpararade der schnellsten Rechner der Welt», Die Zeit, 13 June 1997.

50 Jack J. Dongarra, James R. Bunch, et al., LINPACK Users' Guide, Philadelphia 1979, pp. 803–820.

The Top500 list also showed that it was possible to tear down existing structures and mechanisms of established controls, replacing both with more complex, yet flexible rules. Naturally, machines that were on the Top500 list, and certainly the ones occupying the first ranks, belonged to a supercomputer elite that participated in a competition at the limit of computability. However, these positions could be lost within a few months. Because the ranking came out twice a year – announced in June at the International Supercomputing Conference in Mannheim (the successor to the Mannheim Supercomputer Seminar) and in November by the IEEE Supercomputer Conference in the US – there was no long-term guarantee for a ranking position.[51]

Under a regime of such visible dynamism in HPC, neither the rules of distribution of funds in the 1970s nor the courageous claims of regional sovereignty that later carried the day for a relatively short period could ultimately prevail. Subsequently, a new form of digital federalism emerged – a regime in which a broad variety of actors and programs participated. There was no predefined model and no organizational standard for supercomputing in Germany during this period. Each supercomputer installation represented a careful reconfiguration that met local requirements and somehow fulfilled the federal requirements of participation, competition, and distribution of opportunities in a unique technopolitical cluster of interests. No local *spiritus rector*, no powerful federal institution, no malicious lobbyist, and certainly no powerful hardware provider could have invented such a scenario. The rather unlikely configurations that materialized over time had to be constructed in three different trading zones, involving substantial experimental risk and a considerable amount of organizational creativity.

The first trading zone had to deal with organizational questions of HPC in the budgetary context of a university. Typically, universities monitored their annual budgets exclusively on the basis of expenditure. Long-term investments were difficult to represent in their accounting systems; and income made everyone nervous if it stemmed from any source other than governmental budget allocations or tuition fees. A university was an institution that knew how to spend public money, not an enterprise geared to in-

51 Jack J. Dongarra, Piotr Luszczek, Top500, in: David Padua (ed.), Encyclopedia of Parallel Computing, Boston 2011, pp. 2055–2057, on p. 2056.

come, not to mention profit. However, buying and operating a supercomputer did indeed mean a huge investment, and the University of Stuttgart was thus forced to contemplate selling a share of the computing time on its Cray-2 system. University administrators had to learn what kind of «business» a supercomputer represented. How should the university finance an investment ahead of demand? Was there a reasonable amortization rate for conspicuous computing? How could operating costs be justified independent of actual demand? Was it possible to develop differentiated tariffs for different users, and what did that mean for the university's own users or scattered potential users from other universities in Baden-Württemberg? What did it mean for a power user from another member state, or even from a private enterprise? The answers to such questions had a major impact on the financial governance of the university. An appropriate estimation of costs was a condition *sine qua non*, and this estimation had a deluge of consequences for the existing system of subsidies.[52]

During the early 1990s, after many rounds of intensive discussions and difficult negotiations, the University of Stuttgart found quite an inventive (and complicated) solution. Investment and service (as measured in computing time) should evolve in separate organizations. Founded in 1995, hww (Höchstleistungsrechnen für Wissenschaft und Wirtschaft) GmbH was a specialized company for HPC for science and business that assumed the task of financing future supercomputing investments. The company was set up according to a shareholder model in which Baden-Württemberg, the University of Stuttgart, debis Systemhouse (an IT service provider owned by Daimler), and Porsche participated with different shares.[53] One year later, in 1996,

52 See UASt, Jürgen Blum, Vorwort, in: Blum, Höchstleistungsrechner, pp. 2–3.

53 The University of Stuttgart held 25 percent of the share capital, the State of Baden-Württemberg also 25 percent, debis Systemhaus GmbH 40 percent, and Porsche the remaining 10 percent. Cf. UASt, Rechenschaftsbericht des Rektors Prof. Dr. Heide Ziegler, 1. Okt. 1994 – 30. Sept. 1995, p. 9. Two conditions were attached to the cooperation model. First, participation should be open to other universities – the University of Karlsruhe was treated carefully in this respect by the Baden-Württemberg government. Second, parity between industry and science should always be maintained in the participation model; each side should have a maximum of 50 percent of the share capital. The chairmen of the shareholders' advisory board, appointed by the state of Baden-Württemberg,

the university founded the Höchstleistungsrechenzentrum der Universität Stuttgart (High Performance Computing Center Stuttgart, HLRS). This organizational entity took care of allocating computing time to German universities, specialized research institutes, and industrial users. It is worth noting that the HLRS allocated computing time based on a federal scheme of subsidies.[54] Its 12-member controlling board regulated the allocation of time by approving or dismissing applications for supercomputing services. The board also recommended the acquisition of additional hardware and software. Half of the board members were appointed by the DFG, and the rector's conference of Baden-Württemberg sent in the others.[55] This model finally provided enough federal flexibility for the prevailing legal and accounting rules (venturing to «the edge of legality», as Stuttgart's university chancellor had urged back in 1986); and it furnished a new mode of cooperation between science and the global industrial players situated in and around Stuttgart.

A second trading zone emerged around the problem of providing Germany's research landscape with supercomputing power through a network

were responsible for monitoring this rule. Cf. Ministerium für Wissenschaft und Forschung Baden-Württemberg, Informations- und Kommunikationstechnik in den Hochschulen des Landes Baden-Württemberg, Ausstattungsplan für den Zeitraum 1995 bis zum Jahr 2000 (EDV-Gesamtplan IV), Stuttgart 1995, p. 102.

54 Harms and Meuer provide a concise overview of the model: «The idea is to think big: the state has firmly committed its share of 15 million marks and another 20 million marks for a second tranche. The federal government will contribute supplementary funds of 15 million marks. The Science Council has already endorsed the federal funding of 20 million marks. Industry will contribute appropriate computers and computer use worth more than 40 million marks, as well as associated know-how, to the cooperative effort, i.e., 70 million marks and a total of 110 million marks in the final stage for the ‹super center› in Stuttgart.» Uwe Harms, Hans-Werner Meuer, Höchstleistungsrechnen in Deutschland – ein Rückblick, in: PIK, 1995, 18(2): 100–107, on p. 106.

55 Höchstleistungsrechenzentrum Stuttgart (HLRS), Richtlinien für die Organisation, die Nutzung und den Betrieb, 14/6/1996, Stuttgart 1996. Just under 50 percent of the total capacity was to be reserved for users from the Federal Republic of Germany, about 30 percent for the universities in Baden-Württemberg, and about 20 percent for local demand at the University of Stuttgart. Only 8 percent of the total capacity was to be made available to industry. For industry, debis Systemhaus was to organize the distribution of the capacity, while Porsche wanted to use the computers exclusively for its own needs.

of high-speed data connections. The German Council of Science and Humanities started the discussions and published two reports in 1995.[56] These reports were an institutional answer to the Top500 ranking. The Top500 ranking was a convenient instrument for judging Germany's national supercomputing competitiveness, especially for administrators who lacked expertise in the technical and operational details of the field. Reading the Top500 list in search of a prominent German supercomputing center, however, returned a devastating result. Germany not only ranked way behind Japan and the US, but also behind France, Great Britain, Italy, Switzerland, Canada, and Korea. It was, so to speak, not on the map of global supercomputing.[57] Buying a single new machine could reverse the devastation and restore «Germany» to the ranks of the top players in the field. This was the reason the Science Council wished only to deal with future centers at the highest level of supercomputing.[58]

It was clear from the Top500 list that such a center did not actually exist in 1995. However, the Science Council had no intention of developing a new funding scheme. Rather, it called for «competition among the supercomputing centers».[59] The competition would result in a few exclusive centers providing the necessary supercomputer capacity for science and research.[60] In order to avoid the creation of regional principalities – the Science Council vividly remembered Lothar Späth's act of regional independence (or disobedience) – the council formulated yet another condition for the future development: from then onward, supercomputers would serve all universities in Germany and would be financed by one or various member states.[61]

For the German HPC community, the Science Council overshot the mark – its approach involved too much politics and at the same time too

56 Wissenschaftsrat, Empfehlungen zur Bereitstellung leistungsfähiger Kommunikationsnetze für die Wissenschaft, Saarbrücken 1995; Wissenschaftsrat, Empfehlung zur Versorgung von Wissenschaft und Forschung mit Höchstleistungsrechenkapazität, Kiel 1995.

57 Wissenschaftsrat, Empfehlung, p. 9.

58 Ibid., p. 17.

59 Ibid., p. 24.

60 Ibid., p. 17.

61 Ibid.

much focus on hardware performance. Nonetheless, the HPC community knew that it had to become active in politics if it wanted to shape the development in a politically subtle and somehow more science-oriented way. The community found an interesting and interested ally in the recently restructured Federal Ministry for Education and Research. In October 1997, the directors of Germany's most powerful supercomputing centers came up with a feasibility study for developing national supercomputing. In contrast to the Science Council's visions of competition, the directors proposed a cooperative mechanism. In their eyes, cooperation was the key to propelling Germany to an internationally competitive level. Instead of a destructive arms race in hardware acquisition, the feasibility study suggested focusing on networks and answering the question of how German universities might get access to the future national centers. The possibility of an organizational network of centers of excellence as well as of networking the universities was the response to the Science Council's strategy of competition. In this trading zone, too, new types of areas of negotiation emerged between the federal government, the federal states, and the universities and their supercomputing.

Finally, a third trading zone dealt with promising supercomputing architecture. The concept of massively parallel computing was not new. Initial attempts date back to the 1970s and the famous Illiac IV developed at the University of Illinois.[62] However, it was only in the late 1980s that parallel computing became a useful alternative for Cray's vector-based supercomputing. Massively parallel systems came with many relatively cheap interconnected microprocessors and a distributed memory concept. The promise was an improved price-performance ratio; the problem was coming up with adequate programming techniques.[63] Neither the promise nor the problem was easy to assess. Some people familiar with the field maintained that workstation clusters might soon form a third way. Was this perhaps the end of conventional vector architecture for the entire, and in fact very heterogeneous, user community of supercomputers? Or was parallel computing just a spe-

62 R. Michael Hord, The Illiac IV: The First Supercomputer, Berlin 1982.

63 Richard A. Jenkins, Supercomputers of Today and Tomorrow: The Parallel Processing Revolution, Blueridge Summit, PA 1986.

cialized architecture for computational fluid dynamics in engineering and meteorology? Would further theoretical work eventually show that applied mathematics or theoretical chemistry and particle physics might also benefit from the new «weapon of mass computation»?[64]

The architecture trading zone became even more active when Stuttgart began coupling a brand-new vector computer (NEC SX4) and a powerful massively parallel system (Cray 3TE). Both machines made it into the Top500 list.[65] Now was the time to develop software tools that enabled communication between heterogeneous local systems.[66] The concept of connecting a national supercomputer offering the highest performance, as laid out in the 1997 feasibility study, transformed the architecture problem into a connection problem. Moreover, it served as an indicator that the local, national, and (most likely soon) international diversity of different systems was something to be expected and exploited. This development was the main reason behind Stuttgart's participation in a spectacular metacomputing experiment (with Pittsburgh and Sandia, in Albuquerque, New Mexico) and its efforts to expand supercomputing beyond the dominant institutional and political limits. Participants remember these experiments as «heroic, but unsuccessful». For the Supercomputing '97 conference, Stuttgart and its partners in the US constructed a virtual computer with 1024 processors connected through an extremely heterogeneous network. From the supercomputing center in Pittsburgh, this connection led through the «very high-speed Backbone Network Service (vBNS)» and STAR TAP, both NSF-run projects, to the Canada-based CANARIE and Teleglobe networks right into the network of Deutsche

64 In 1984, the journal *Parallel Computing* started to appear. For the German discussion, see Hans-Werner Meuer (ed.), Parallelisierung komplexer Probleme. Einsatz von Parallelrechnern in Forschung und Industrie, Berlin 1991; Robert Ahlrichs, Parallelrechner versus Workstation Cluster. Positionspapier, in: Hans-Werner Meuer (ed.), Supercomputer '93. Anwendungen, Architekturen, Trends. Mannheim 24.–26. Juni 1993. Proceedings, Berlin 1990, pp. 179–180, on p. 179.

65 In June 1996, the NEC SX-4/32 was number ten in the Top 500 list. In June 1997, Stuttgart's T3E Cray HPE was number ten, while the NEC SX-4/32 had already descended to rank 38.

66 UASt, Michael Resch et al., PACX-MPI, in: Informationen für Nutzer des Rechenzentrums (Heft 11/12) 1997, pp. 13–14.

Telekom AG, and from there to Stuttgart. Obviously, data also found its way back to Pittsburgh.[67] While the experiment could not overcome latency problems, Stuttgart acquired in Sandia a partner highly skilled in both wide area network technologies and computer-based methods for visualization. At the same time, Stuttgart was clearly the go-to location for any questions regarding grid computing in Germany.[68]

Conclusion

For scholars in public law it is no surprise that a developing federalism implies many corresponding differentiations for local and regional entities.[69] When the hotspots of political responsibility moved to federal actors, local actors had to develop matching structures – and vice versa. This holds true for digital federalism as well. When member states started to play a more competitive game in supercomputing (as they did in the mid-1980s), the federal system of technoscientific policies had to adjust its funding programs for the enhancement of computing, its policies for the development of universities, and its support for specialized research institutes. Supercomputing is such an expensive endeavor that no political entity, program, enterprise, or university is capable of acting alone. Some of Germany's cutting-edge facilities in supercomputing had to learn that lesson toward the end of the 1990s. It became clear that building and running a supercomputing center necessarily means cooperating selectively and excluding efficiently at the highest level of performance. In other words, success in supercomputing is extremely unlikely and requires the simultaneous development of rules of configuration

67 Pittsburgh and Stuttgart Inaugurate High-Speed Transatlantic Metacomputing, Pittsburgh Supercomputing Center, Press Release, 24 June 1997.

68 Thomas J. Pratt et al., Sandia's Network for SC '97: Supporting Visualization, Distributed Cluster Computing, and Production Data Networking with a Wide Area High Performance Parallel Asynchronous Transfer Mode (ATM) Network, in: Sandia National Laboratories (ed.), Sandia Report SAND98-1154, Albuquerque 1998.

69 Dietrich Schindler, Schweizerischer und europäischer Föderalismus, in: Schweizerisches Zentralblatt für Staats- und Verwaltungsrecht, 1992, 93(5): 193–223.

for machines, programs, users, and sponsors.[70] Hence, supercomputing requires strong alliances, carefully designed forms of autonomy, and selective interrupts for the control of desired and disruptive interdependence. Supercomputing's configuration is very much akin to a confederation. No wonder the multiple trading zones of supercomputing have formed the primary playing ground for emerging digital federalism since the late 20th century.

This conclusion is by no means valid only for supercomputing and its trading zones. The transfer of administrative work (both official and private) into the personal computer, and the concomitant use of local and wide area networks by interacting bureaucracies have played an equally important role in the development of digital federalism. However, arriving at this insight implies at least three analytical preconditions. First, having the courage to distinguish between different types of computing machines, notwithstanding claims of universality. Second, overcoming the pseudo-critical difference between the center and the periphery, and instead recognizing the entanglement of sociotechnical circumstances. Finally, abandoning the search for a hidden masterplan conceived by a *spiritus rector*, a powerful interest group, the *Zeitgeist*, the counterculture, or any other conspiracy to explain the messy, yet powerful situation in which digital federalism finds itself at the end of four decades.

70 On the age of configuration, see David Gugerli and Magaly Tornay, Das Zeitalter der Konfigurationen 1980 bis 2010. Ein Beitrag zur zeithistorischen Debatte, in: Historische Anthropologie, 2018, 26(2): 224–244.

«A story of friendship and misunderstandings»: the origins of the Swiss National Supercomputing Centre 1985–1992

Paolo Bory, Ely Lüthi, Gabriele Balbi

Abstract

This article retraces the origins of the Swiss National Supercomputing Centre (CSCS), founded in Manno in 1991, and the fervent debate surrounding them. Relying on a vast corpus of primary sources collected in institutional, private, and national archives, we show how the early history of CSCS reflects key features of the Swiss federalist model, while also revealing its ambitions, challenges, and contradictions. The two main questions the research sought to answer are: Why build a supercomputer center in Switzerland? And why in Ticino? Our research shows that, as was commonly noted at the time, the supercomputer signaled a further step toward the digitization of Switzerland and was an act of solidarity with the Italian-speaking area of the country. Moreover, the supercomputer was intended to enhance Switzerland's international dimension, stimulating collaborations with other European countries and even beyond. Finally, by highlighting the longstanding tension between centralizing and decentralizing efforts, the early history of CSCS reveals how, rather than following a linear trajectory, digital federalism faced a series of difficult political, economic, and cultural choices.

Introduction: a history that «must be written down»

On 2 December 1985, the Swiss Federal Council decided to apply special measures in favor of training and research in computer science and engineering.[1] One outcome of these measures was the founding of the Swiss National Supercomputing Centre (CSCS), today one of the world's most important centers in this field.

1 Swiss Federal Chancellery, Botschaft über Sondermassnahmen zugunsten der Ausbildung und Weiterbildung sowie der Forschung in der Informatik und den Ingenieurwissenschaften vom 2. Dezember 1985 [Message Concerning Special Measures in Favor of Training and Further Education as well as Research in Computer Science and Engineering of 2 December 1985], in: Bundesblatt 1986, 1(5): 321–383.

Relying on archival research and an extensive corpus of primary sources, this article retraces the path that led to the founding of CSCS in Manno (canton of Ticino), in the Italian-speaking part of Switzerland. As we will show, even more than the center's founding, the decision about where to locate it triggered a political debate among Swiss institutions and stakeholders from 1985 up to 1992, when CSCS was inaugurated. In fact, in the late 1980s and early 1990s, the canton of Ticino still lacked scientific and academic institutions.

During his inaugural speech in 1992, States Council member Giuseppe Buffi, a key figure in national and local politics at the time, argued:

> The choice of the building in which to locate the center was a matter of controversy, of decisions taken then reneged on, of misunderstandings: this is the reason why I said [at the start of this speech] that this is a story of friendship and misunderstandings. [...] I think that one day, a detailed history of the awarding of CSCS to Ticino must be written down, so as to provide an illustrative example of what can happen in our local reality.[2]

This article is a first attempt to fulfill Buffi's wishes; but it is also something more. As we will show, this story reveals key features of the Swiss federalist model, including its ambitions, difficulties, contradictions, and the complex geographic, political, and scientific relations that the model itself entails. Retracing the pre- and early history of CSCS is thus essential to understanding how different actors imagined the digital future of Switzerland and the way in which the founding of a federal institution was negotiated among these actors. Our main goal is to elucidate the relationship between the history of CSCS and the process through which the location of a supercomputer in the Italian-speaking part of Switzerland gained consensus in the late 1980s to early 1990s.

2 CSCS Archive, 1.10.1992 Inaugurazione CSCS, For the press, Giuseppe Buffi's inaugural speech, pp. 5–6. Original text in Italian: «La scelta dell'edificio dove dare sede al centro fu motivo di polemiche, di decisioni prese e rimangiate, di equivoci: ed è questa la ragione per cui ho affermato che questa è una storia di amicizie ed incomprensioni. Penso allora che un giorno la storia dell'assegnazione al Ticino del Centro Svizzero di Calcolo Scientifico andrà scritta nei minimi particolari, perché rimanga testimonianza emblematica di quanto può capitare nella nostra realtà locale.»

The paper is organized in four sections. The first section outlines the reasons for studying supercomputing history, a still uncharted subject in media and technology studies. In the second section, we summarize the archival work and the corpus of sources selected for analysis. The third section is dedicated to analyzing the main theoretical and political issues surrounding the birth of CSCS, and especially its location in the Italian-speaking part of Switzerland: the decentralization efforts of the Swiss Federal Institute of Technology in Zurich (ETHZ), the Federal Council and the federal government; the troubled path to and the local dispute regarding the center's location; and the future of CSCS as a bridge to international relations. The last section, in addition to giving a brief nod to the unsolved issues that further research should address, summarizes how the origins of CSCS can be read vis-à-vis digital federalism, stressing three key concepts that characterize this historic case: CSCS as a bridge, as a decentralization strategy, and as an act of solidarity.

Supercomputers in media and technology studies

The history of supercomputers constitutes a blind spot in media and technology studies. In these fields, from the perspective of the humanities and social sciences, computers are depicted as sociotechnical machines with a four-stage history. The first, called the mechanical age, goes from 2500 BC to the 1930s, with intensification from the 19th century when the birth of the nation-state, the industrial revolution, and the emergence of mass society fueled demand for calculation. During a second age, called the mainframe age, which ranged from the mid-1930s to the mid-1970s, the military and large corporations built powerful computers to fulfill their needs. The third age, from the mid-1970s to the 2000s, saw the rise of small computers bought by ordinary people that were also used for everyday needs. Finally, in the last 20 years – a post-PC age – other digital devices, such as smartphones and tablets, have complemented PCs.[3] In general, supercomputers are not included in this conventional story.[4]

Nevertheless, studying supercomputers and their histories is more and more relevant for media and technology studies from a political, economic, technical, and sociocultural perspective.

The scientific literature on supercomputers comes mainly from the hard sciences, such as informatics, computer science, mathematics, physics, and chemistry. This literature shows the relevance and application of supercomputers in the respective fields: supercomputers are powerful tools for doing numerical simulations and analyzing huge amounts of data; or they are useful in scientific and industrial research and in the military.

In general, supercomputer histories are just ancillary parts of this literature. These histories focus primarily on techno-nationalist battles between the United States and Japan, the two most important producers of supercomputers up to the 2000s, when China started to become the global leader in the field. The United States and Japan fought for supremacy in the industry through two companies that adopted opposing paradigms: Cray, the American champion of parallel supercomputing, and Japan's NEC, which produced vector supercomputers.[5] Supercomputers have always been linked to national and, to an extent, nationalistic visions. For example, even today supercomputers continue to compete based on calculating power in a race called the Top500 project, which ranks the 500 most powerful non-distributed computer systems in the world.[6]

3 Gabriele Balbi, Paolo Magaudda, A History of Digital Media: An Intermedia and Global Perspective, New York: 2018, pp. 27–66.

4 For a compelling analysis of the history of supercomputing as conspicuous technology, see also David Gugerli and Ricky Wichum's article in this issue.

5 See for example, Marie Anchordoguy, Japanese-American Trade Conflict and Supercomputers, in: Political Science Quarterly, 1994, 109(1): 35–80; Susan L. Graham et al., Getting Up to Speed: The Future of Supercomputing, Washington DC 2005, Ch. 3; Yoshio Oyanagi, Development of Supercomputers in Japan: Hardware and Software, in: Parallel Computing, 1999, 25: 1545–1567.

6 In November 2020, the first ten supercomputers were listed in the United States (4), China (2), Japan (1) – the most powerful supercomputer in the world – Italy (1), Germany (1), and Saudi Arabia (1). CSCS slipped from the top ten in 2020. Looking at the

But supercomputers are not only about computing power. They also matter in terms of the political economy of communication, a discipline that deals with the distribution of power and the role of states and private companies in media and communication markets.[7] Supercomputers are tools supported by national strategies and public money in areas such as telecommunications, broadcasting, the movie industry, and other media. National (and international) armies, private companies, and research centers use supercomputers for different purposes.

However, despite their relevance for national governments and businesses, why and how supercomputers influence people's everyday lives is not well known. Media and technology studies focus on how consumer goods are incorporated in people's lives – newspapers, books, radios and TVs, audio devices, movies, comics, cameras, smartphones, PCs, and dozens of other devices that are bought and used by people at home or on the move for information and entertainment. Supercomputers are not «visible» and, apparently, not relevant to everyday life. They are not visible because they are generally hosted in closed spaces, sometimes far from city centers, and only a limited number of people know of their existence. Supercomputer infrastructures are hidden as well, despite their huge size: a supercomputer needs cooling systems (water or air-conditioning), big hangar-like buildings to accommodate humming machines, electrical energy grids, and high-speed digital networks to receive and send data before and after calculation.[8] This mostly

distribution of supercomputers by country, China is first (214, 42.8 % of the total system share), the United States second (113, 22.6 %), and Japan third (34, 6.8 %). Source: https://www.top500.org/statistics/list/ (23/11/2020).

7 The literature on the political economy of communication is extensive, like the «classics» in the field. For a recent and updated reconstruction, see Dwayne Winseck, Reconstructing the Political Economy of Communication for the Digital Media Age, in: Political Economy of Communication, 2016, 4(2): 73–114.

8 On the material dimension of supercomputers, see Thomas A. Weber, The National Science Foundation Supercomputer Centres Program, in: International Journal of Supercomputing Applications, 1991, 5(4): 3; Tapasya Patki et al., Supercomputing Centres and Electricity Service Providers: A Geographically Distributed Perspective on Demand Management in Europe and the United States, in: Michela Taufer et al. (eds.), High Performance Computing. ISC High Performance 2016. Lecture Notes in Computer Science,

invisible material dimension, a topic which has mainly been studied by science and technology studies,[9] represents a second approach to studying supercomputing that focuses on different and combined infrastructures and stakeholders at the national and transnational level.

Supercomputers are invisible not only because of their hidden infrastructures but also because they seem to be useless for citizens. This impression is incorrect, but, again, it is difficult to explain the role of supercomputers in everyday life. A media studies approach to supercomputing should be helpful in this sense. John E. Aldag claims that «supercomputers have the potential to have a broad, pervasive, and very positive impact in the world around us».[10] This does not mean that «normal» people use or will use them (academics and industrial researchers are and will probably continue to be the main users), but people keep on benefiting from supercomputer calculations. The automotive, oil, nuclear, molecular, health, and many other sectors use supercomputers for simulations, helping to improve security standards and to drive innovation. In the nuclear sector, for example, instead of carrying out destructive and socially unpopular tests, supercomputers can simulate them. In the oil industry and in the field of atmospheric pollution, supercomputers are able to reproduce via longitudinal simulations what would take decades to observe. Other very popular fields that require huge

Cham 2016; Mario Romero et al., Supercomputers Keeping People Warm in the Winter, Paper presented at the 2nd International Conference on ICT for Sustainability (ICTS), 24–27 August 2014, Stockholm, http://urn.kb.se/resolve?urn=urn:nbn:se:kth:diva-159065 (23/11/2020). See also Guido Scherrer, Manno: A Superbrain for Switzerland, in: Monika Dommann et al. (eds.), Data Centers: Edges of a Wired Nation, Zurich 2020, pp. 30–44.

9 In the last decade, media studies and science and technology studies have constantly overlapped and influenced each other. In order to explore this disciplinary connection, see, for example, Pablo Boczkowski and Leah A. Lievrouw, Bridging STS and Communication Studies: Scholarship on Media and Information Technologies, in: Edward J. Hakkett et al. (eds.), The Handbook of Science and Technology Studies, 3rd Edition, Cambridge MA 2008, pp. 949–977.

10 John E. Aldag, The Impact of Supercomputers: Global, Pervasive, Positive, in: International Journal of Supercomputing Applications, 1989, 3(2): 3.

calculations include climate change[11] and weather forecasting. For example, people get forecasts on their smartphones without asking where they come from or who produced them. Supercomputers make these forecasts possible through complex calculations that only powerful machines can do. Other examples exist. But the point is that supercomputers play a crucial role in routine areas of modern society. Supercomputers enable these practices, and for this reason studying them can help to better understand contemporary social practices and innovation.

Archival research and sources

To investigate the origins of CSCS, we mainly selected sources from two archives: the CSCS archive in Lugano and the ETHZ archives in Zurich.[12] We searched primarily for documents from 1985 to 2000, although we also found and collected previous and later relevant materials. The physical visits were followed by online research to find missing documents, for example, some of CSCS's annual reports. In total we have collected and digitalized a total of over 8000 documents.

These sources are diverse and are mainly written in German, although they also include French, Italian, and English documents. Projects to build a national supercomputing center are the most frequent types of documents. Among these, we found various project proposals published by ETHZ, other national institutions and some private companies, reports, and evaluations on the various proposals (mainly published by ETHZ and the Swiss federal government), as well as personal letters and exchanges of communication between the numerous political and economic actors involved (among which

11 Paul N. Edwards, A Vast Machine: Computer Models, Climate Data, and the Politics of Global Warming, New York 2010.

12 The ETH Zurich University Archives mainly contain documents from ETHZ and from the ETH Council. We went through documents from these sources between the years 1985 and 2000, in particular administrative documents from the ETH «Schulleitung» (ETH Executive Board). Access was possible after a request to the General Secretariat of ETH Zurich. Other documents were subject to an embargo period and were therefore not provided in certain cases.

the ETH Council[13]). Other sources included internal material from CSCS and ETHZ, such as magazines, press releases, flyers, and brochures. Finally, materials from national and local political actors, international and Swiss tech-magazines (mostly focused on different types and uses of supercomputing), as well as articles published in Swiss newspapers and programs for public events and research seminars also proved useful.

The most relevant sources were translated into English and categorized according to different political, economic, and even geographic visions. Especially in the ETHZ archives, we found several sources regarding the importance of federalism and decentralization in supercomputing. In the CSCS archive, issues related mainly to location and to political and industry debates on the rationale for the center. Both archives contain sources and documents on future users and on the international connections the CSCS was thought to be able to build. The press release folders were crucial for understanding how Swiss newspapers reported on the origins of the center in the German- and Italian-speaking areas.

At a later stage of this research, we were able to get in touch with Fulvio Caccia, a member of Ticino's Council of State from 1977 to 1987 and of the Swiss National Council between 1987 and 1998. Caccia provided us with materials from his personal archive, collected during the four-year mandate assigned to him by the Ticino State Council in 1987 to develop the cantonal university, in particular possibilities for collaboration with the universities and the two Swiss Federal Institutes of Technology.[14] These materials comprise several personal letters and a memorandum in which Caccia explains his role in the founding of CSCS.

13 The Federal Council had mandated ETHZ to implement the supercomputing project. However, because it was a large national project, decisions needed to follow a specific path. Decisions were first taken by the ETHZ Executive Board, then by the ETH Council (which involved representatives of both ETHZ and EPFL), and then finally the Federal Council. For this reason, we also examined documents involving the ETH Council and the Federal Council.

14 Fulvio Caccia, personal collection, Memorandum on CSCS, 30/9/1991.

Why build a supercomputing center for Switzerland?

An «act of solidarity» by ETHZ: a supercomputer to promote informatics in Switzerland

The federal government's decision of 1985 to foster the development of digital infrastructures in Switzerland not only resulted in a supercomputing center; it also gave rise to a national university network named SWITCH.[15] The aim of SWITCH was to provide Switzerland with a decentralized infrastructure connecting different research institutions, whereas the supercomputing «center» would act as a key hub in which raw data would be collected and processed for scientific purposes. The Swiss Federal Council had several motives for acquiring a scientific supercomputer and building a computer network among universities. The «information society» was «at the door» and «no one» could «stop the technological revolution».[16] According to the Federal Council, in the late 1970s and early 1980s Switzerland had «fallen into a new and dangerous dependence on foreign resources».[17] To safeguard its independence, the country had to react. Huge efforts were undertaken at both the technological and intellectual level to educate new engineers, to help Swiss companies compete in the international market, to train new professionals in the field of informatics, and to connect Swiss higher education and research institutions both nationally and internationally. The government's decision was thus in line with global investments in the information and communications technology (ICT) sector and with the mainstream political economy of communication, which encouraged embracing the «information society» in Western countries. As media historians have shown, this idea had

15 For a compelling analysis of the history of SWITCH, see Daniela Zetti's article in this issue.

16 Swiss Federal Chancellery, 1986, p. 273. Original text in Italian: «[È] incontestabile che nessuno è in grado di contenere la rivoluzione tecnologica. La società informatica è alle porte».

17 Ibid. Original text in Italian: «La Svizzera è caduta in una nuova e pericolosa dipendenza dall'estero La Svizzera, paese povero in materie prime, deve già da tempo sfruttare l'unica risorsa di cui dispone, segnatamente la propria ‹materia grigia›.»

been circulating globally at least from the 1970s, if not before, and it would characterize the final decades of the 20th century.[18]

The Federal Council decided to fund the supercomputing project – to establish centralized supercomputing somewhere in Switzerland – in the amount of 40 million Swiss francs and directed ETHZ to implement it.[19] For its part, ETHZ supplemented the public funds with 8 million francs, plus human and technical resources. Such numbers turned out to be only a minimal part of the real investment in CSCS; in the years that followed, managing the supercomputing center would entail considerable annual costs for human and technical resources.[20]

The relevance of CSCS for the future of Switzerland as a nation was evident in its full name: Swiss National Supercomputing Centre. Despite the fact that supercomputing power was already being exploited by Swiss institutions such as the two Swiss Federal Institutes of Technology in Zurich (ETHZ) and in Lausanne (EPFL), the new center represented an innovative service for Switzerland far beyond its cantonal borders. In fact, notwithstanding the key role of ETHZ both in the founding and operation of CSCS, the center was and still is the «Swiss» supercomputing center. In other words, CSCS was (and is) a central national service and facility aimed at accelerating and favoring the digital growth of the country. As stated in the federal

18 Some key national reports and studies from the 1960s to the 1970s have been analyzed in depth in the literature of media and technology studies. Consider, for instance, the influence of Nora and Minc's *Computerization of Society* or Daniel Bell's neologism «post-industrial society». However, historians have paid less attention to certain key texts on the managerial and business aspects of computing, such as articles and reports released by McKinsey and Company. These texts are often cited in the business literature on computing and supercomputing, and they constantly come up in our corpus of sources from the CSCS Archive: see, for example, Simon Nora and Alain Minc, L'informatisation de la société, La documentation française, Paris 1978; Daniel Bell, The Coming of Post-Industrial Society: A Venture in Social Forecasting, New York 1983; McKinsey and Company, Unlocking the Computer's Profit Potential, in: Computers and Automation, 1969, 16(7): 24–33.

19 Fulvio Caccia's Personal Archive, «Memorandum on CSCS», 30/9/1991.

20 ETHZ Archives, Box 12_ER-GS-Leh01:1.5.21f, CSCS Manno & POL-HLR 91_2. Most of this information is contained in a letter from Seehars to Scheidegger on the settlement of the Federal Council's loan for the high-performance supercomputer, pp. 2–8.

dispatch of 1985, the first goal in acquiring the supercomputer was to provide «all» Swiss institutions with «a real computer service able to manage a decentralized system for a heterogeneous usership».[21]

Decentralization, both at the technical and symbolic level, is a crucial concept for the birth of CSCS, especially in terms of national policies and sociotechnical imaginaries.[22] Arguably, decentralization is relevant for at least three reasons. First, it is part of the «digital federalism» model that characterizes Switzerland. In combination with the future network SWITCH, the future CSCS was imagined as a technical tool for decentralizing and, in turn, federalizing Swiss digital resources and services. The idea was to create a new and (partially) independent institution that would provide a kind of «public service» separate from the internal activities of ETHZ and EPFL. Notably, ETHZ played a key role in this process since it was mandated by the Federal Council to implement the supercomputer project and contributed to it with additional funding to create the center. ETHZ has also consistently supervised CSCS workflow, productivity, and budgets.

For ETHZ, locating the center in Ticino meant clearly separating its internal projects – which were kept in Zurich – from its external ones, which were shifted to CSCS. Furthermore, from a more pragmatic perspective, the lack of physical space in the buildings owned by ETHZ and the location in a «neutral» area like Ticino encouraged the support of the governing body of the Swiss Federal Institutes of Technology (ETH Council), thus overcoming the competition between ETHZ and EPFL.

Second, the new center also meant extending computational resources to scientists throughout the country. The supercomputing center was expected to provide all Swiss scientists, engineers, and computer scientists with new tools and professional skills. As stated in a 1990 ETHZ press release:

21 Swiss Federal Chancellery, 1986, p. 283.

22 We use the expression *sociotechnical imaginaries* as defined by Jasanoff and Kim, namely, «collectively imagined forms of social life and social order reflected in the design and fulfilment of nation-specific scientific and/or technological projects». See Sheila Jasanoff and Sang-Hyun Kim, Containing the Atom: Sociotechnical Imaginaries and Nuclear Power in the United States and South Korea, in: Minerva, 2009, 47: 120.

> Now a single national computer is to be added for selected top needs of all universities. The various research groups – above all chemists, physicists, biologists, and, to an increasing extent, engineers – will use this national high-performance computer mostly via fast data lines from their place of work, but in some cases also locally at the location of the computer center.[23]

Actually, digital literacy and specialization in informatics were two milestones for Swiss digital development, increasing access to the «information society» and becoming up-to-date:

> All these measures by the Confederation clearly document the strong determination of the federal authorities to provide Swiss research with the necessary means to keep abreast of the rapid worldwide scientific advances and to develop as much as possible the creativity of its scientists.[24]

Finally, centralizing the computing facilities in Ticino served to deliver a critical facility to a canton that had previously lacked academic and scientific institutions. This decision aimed somehow to balance the gap between the German- and French-speaking parts of the country with the Italian-speaking region. Hence, the decision to locate the center in Ticino, apart from its local impact, was in line with the federalist principle to distribute not only political power but also scientific and infrastructural functions and institutions to bring all the regions closer together. ETHZ president Jakob Nüesch defined the founding of CSCS in Ticino as «an act of solidarity to a promising region for both research and production […] (with) great capacity potential».[25] This sort of *political federalism* also included «the willingness to involve the

---

23 ETHZ Archives, Manno_Akz2000-01_Auswahl, 1989-1991_POL HLR-91_3, ETHZ Press release, 26-27/6/1990, p. 1. Original text in German: «Nun soll ein einziger nationaler Rechner für ausgewählte Spitzenbedürfnisse aller Hochschulen dazukommen. Die verschiedenen Forschungsgruppen – vor allem Chemiker, Physiker, Biologen und in zunehmendem Mass auch Ingenieure – werden diesen nationalen Hochleistungsrechner meist über schnelle Datenleitungen von ihrem Arbeitsort aus benützen, teilweise aber auch lokal am Standort des Rechenzentrums.»

24 CSCS Archive, Project HLR-91, Project basis/initialization, On the initiative for an HSC center in Switzerland, pp. 5-6.

25 CSCS Archive, 1.10.1992 Inaugurazione CSCS, For the press, Nüesch's discourse for CSCS inauguration, pp. 3-4.

Italian-speaking area of Switzerland in the academic sphere through an important national project».[26] Moreover, it went hand in hand with the idea of *social federalism*, which aimed at making CSCS into «an excellent example of a high-level scientific service placed at the disposal of the whole [national] scientific community»,[27] able to «act as a bridge»[28] connecting the higher education sector with national academic and industrial research.

In line with this narrative, Werner Carobbio, member of Ticino's parliament, stated during a parliamentary hearing in 1989:

> The decision to install the new supercomputer for ETH Zurich in Ticino is very welcome, as it recognizes regional policy requirements and the administration's decentralization efforts. In addition, it goes a long way toward enabling the peripheral region of Ticino to move closer to the centers of Zurich and Bern.[29]

Notwithstanding the clear positive stance toward the founding of a «decentralized» center, Carobbio's statement highlights an unsolved contradiction in CSCS history: the fact that the supercomputing center was «for ETH Zurich» as a federal institution and not for Switzerland. Even today, CSCS is a national service and represents the expression of the federalist vision of a decentralized distribution of power; but it is also a symbol of the power of ETHZ. It is a kind of scientific (but also political) «soft power» exercised by the most prestigious academic institution in Switzerland. This contradiction between federalism and scientific authority, or in other words between decentralization and centralization, deeply impacted CSCS in its first decades,

---

26 CSCS Archive, Project HLR-91, Technical documents, Report on CSCS for Ticino government, p. 1. Original text in Italian: «la volontà di coinvolgere la Svizzera Italiana nella sfera accademica con un importante progetto a carattere nazionale».

27 CSCS Archive, Annual Reports, 1992, p. 7.

28 Ibid., 1999–2000, p. 3, Interview with new director Michele Parinello.

29 CSCS Archive, 1989 Manno – Nationales Hochleistungsrechenzentrum, Parliament, Interpellation of Mr. Carobbio, 14/6/1989, p. 3. Original text in German: «Der Entschied, den neuen Supercomputer für die ETH Zürich im Tessin zu installieren, ist sehr begrüssenswert, trägt er doch die regionalpolitischen Anforderungen und den Dezentralisierungsbemühungen der Verwaltung Rechnung. Ausserdem rückt mit Sicherheit auf diese Weise die Randregion Tessin den Zentren Zürich und Bern etwas näher.»

at least until the rebirth of the center when it was moved from Manno to Lugano in 2012.

In fact, following the 1991 inauguration, the promises of outstanding success for the center failed to materialize. Several interrelated reasons explain why. First, according to Fulvio Caccia, the new president of ETHZ, Jakob Nüesch, did not drive the CSCS project with the enthusiasm and long-term perspective of his predecessor, Hans Bühlmann. During the 1990s, the center went through five different directors, and ETHZ had great difficulty managing and supervising its activities from Zurich. Second, despite the construction of new Swiss digital and transportation infrastructures, the physical distance between Manno and Zurich did not help. It took academics and scientists coming from Zurich several hours to reach the center. And although CSCS and the regional airline Crossair[30] were able to find some agreement, the most efficient hub for getting to Manno was far away in the Italian city of Milan. Third, decisions taken by scientists and academics not affiliated with ETHZ were difficult to supervise. The supercomputing market was changing fast in the 1990s, and both technical and managerial decisions required substantial human, political, and scientific resources that ETHZ could not (or perhaps did not want) to use for the center. Finally, from an infrastructural perspective, data connections for supercomputing were slow and very expensive at the time. Indeed, the supercomputer was not a normal client, and supercomputing was not a mainstream application; it required particular effort, also in terms of data management and infrastructure. These needs called for further federal financial and infrastructural investment that would only come about with time. Overall, as current CSCS director Michele de Lorenzi has argued, «ETHZ had its baby. But there is a difference between giving birth to a baby and raising it».[31]

30 On the deal between CSCS and Crossair, see ETHZ Archives, Box 12_ER-GS-Leh01:1.5.21f, R-GS-Leh01: 1–5–21e-f, p. 34, Mail from Grin to Chapuis, 2/7/1997.

31 Interview with Michele de Lorenzi, 18/12/2019.

A «cultural and economic future» for Ticino: the choice of Manno and the dispute about location

As soon as the Federal Council decided to finance a new supercomputer, discussions started on where to locate it. In the early years, both ETHZ and EPFL wanted to host the supercomputer and to secure the federal funds for it. This tension between the two institutions blocked the decision process for some time. Then, at the beginning of 1988, it was decided to establish a new research center to host the supercomputer (this was not taken for granted in the federal dispatch) and serious location proposals started to come in. The search for a location was also a race against time because, as stated in the federal dispatch, «[the] period of validity [of the fund] is limited to five years and, in view of the difference between the academic year and the calendar year, will begin on 1 October 1986 and expire on 30 September 1991».[32]

A definite location therefore needed to be found as soon as possible, also given the fact that ETHZ was «already planning to install the national HLR [*Hochleistungsrechner*, ‹supercomputer› in German] in mid-1991»; as such, any «postponement of the installation should be avoided».[33] Hence, it was agreed that «the location decision should be finalized in April 1989», so as to allow the center's installation «in a timely manner and without significant inconvenience to users».[34]

In February 1988, as part of his mandate from the Ticino State Council, Fulvio Caccia went with Giuseppe Buffi to meet ETHZ president Hans Bühlmann. When asked about establishing collaborations between ETHZ and Ti-

32 Swiss Federal Chancellery, 1986, pp. 270–271. Original text in Italian: «La durata di validità è limitata a cinque anni e, in considerazione dello sfalsamento tra anno accademico e anno civile, inizierà il 1° ottobre 1986 e scadrà il 30 settembre 1991.»

33 ETHZ Archives, Manno_Akz2000-01_Auswahl, 1989-1991_POL HLR-91_10, p. 3. Original text in German: «[D]ie ETH Zürich plant schon heute mit der Zielsetzung, dass der nationale HLR Mitte 1991 installiert wird. Deswegen soll eine Verschiebung der Installation vermieden werden.»

34 CSCS Archive, Project HLR-91, Realization phase, Bühlmann on the location in Ticino, pp. 1–2. Original text in German: «Ein Standort Tessin ist für den nationalen HLR zeitgerecht und ohne wesentliche Nachteile für die Benützer möglich, sofern der Standortentscheid im April '89 definitiv fällt.»

cino, Bühlmann affirmed that a first step could be the realization in Ticino of a scientific seminar center related to ETHZ (Centro Stefano Franscini at Monte Verità, Ascona); subsequently, a bigger project, such as a supercomputing center, could be evaluated. Seven months later, the seminar center had advanced very well, and so Bühlmann confirmed his interest in locating the supercomputing center in Ticino.[35]

Nevertheless, two main objections emerged at the federal level: the telecommunication networks in Ticino were not equipped to transmit the required volumes of data and, especially, there was not enough time to build the center by September 1991.[36] Buffi then pushed Caccia to do his «best in order to avoid the closure of the dossier, recognizing at the same time that a public building with such special needs could not be built in three years».[37]

At the same time, Silvio Tarchini, a successful local entrepreneur active in the construction industry, was about to erect several industrial buildings in Manno (seven kilometers north of Lugano), at a site he had just rented from the Swiss Federal Railways (SBB) with a lease term of sixty years. He then planned to rent these buildings to various other companies. When Caccia learned about this project, which was already at an advanced stage, he met Tarchini and explained the supercomputing project to him. Interested in taking part, Tarchini proposed to rent ETHZ a space inside one of the SBB buildings to house CSCS.[38] Close to the railways, several industries, and the airport, Manno seemed the perfect location for the center.

In early September 1988, Caccia met with the ETHZ Executive Board. After some initial skepticism,[39] and the ironic observation on the part of some

35 Caccia, Memorandum on CSCS, p. 2.

36 Ibid.

37 Ibid. Original text in Italian: «Buffi m'incarica di fare il possibile per non lasciar chiudere l'incarto in quel modo, pur convenendo che in tre anni non si realizza nessun edificio pubblico con esigenze speciali.»

38 Ibid., p. 3.

39 «With the tact that denials can well justify, I am told clearly that there is nothing to do, that this is not the right opportunity for Ticino, that we need to look for another one». Original text in Italian: «Con la delicatezza che i dinieghi possono ben giustificare, mi si dice in modo chiaro che non c'è niente da fare, che non è l'occasione buona per il Ticino, che occorre cercarne un'altra». Ibid.

that the center risked ending up being a «cathedral in a desert», Caccia persuaded the board to create a working group to «investigate whether and at what price the structural, technical, and personnel infrastructure required for the national high-performance computer could be provided in Ticino» and, if yes, which location would better suit the center.[40] This working group included ETHZ vice-president Carl August Zehnder, his secretary Mr. Schindler, Fiorenzo Scaroni from the ETHZ computing center, and Caccia.[41] The group sent its final report to Bühlmann on 1 March 1989; the report was accepted a week later by the entire ETHZ Executive Board and on 29 March by the ETH Council. The Federal Council was then informed of the Manno site's suitability and, after further evaluation, final approval was given to the project.[42]

The first hurdles to the location were not long in manifesting, however. On 3 June 1989, an article published in the *Tessiner Zeitung*[43] reported that the previous owners of the site in Manno, which «was expropriated from them in the seventies», had «filed a complaint with the court» against the SBB, which was using the site «for purposes other than [those which led to the] expropriation» (Fig. 1). Indeed, the SBB had acquired the land to construct a goods depot and later decided «to rent out the remaining part that was not needed» to several investors, including Silvio Tarchini, who then offered part of the site «to the Federal Office for Buildings [and Logistics] for rent».[44]

40 CSCS Archive, 1989 Manno – Nationales Hochleistungsrechenzentrum, Parliament, Answers to questions on HSC center. Original text in German: «Die Schulleitung der ETH Zürich setzte in September 1988 eine Arbeitsgruppe mit dem Auftrag ein, zu untersuchen, ob und zu welchem Preis im Tessin die für den nationalen Hochleistungrechner erforderliche bauliche, technische und personelle Infrastruktur bereitgestellt werden kann.»

41 Fulvio Caccia, personal communication, 27/8/2020, and Caccia, Memorandum on CSCS.

42 CSCS Archive, 1989 Manno – Nationales Hochleistungsrechenzentrum, On location, Minutes ETH Council, 29/3/1989, and Caccia, Memorandum on CSCS.

43 The *Tessiner Zeitung*, published weekly in Ticino, is a German-speaking newspaper aimed at German speakers living in Ticino.

44 CSCS Archive, 1989 Manno – Nationales Hochleistungsrechenzentrum, Press articles, Tessiner Zeitung, 3/6/1989. Original text in German: «Ihnen war das 140 000 Quadratmeter grosse Grundstück in Manno in den siebziger Jahren enteignet worden. [...] Die Bundesbahnen beschlossen deshalb, den nicht benötigten Rest zu vermieten. [...] Unter den Mietern figuriert auch der Luganeser Bauunternehmer Silvio Tarchini. Er

3. Juni 1989

/3772

Die ehemaligen Besitzer des SBB-Geländes in Manno haben beim Gericht Klage eingereicht

Erste Hindernisse für das geplante Super-Elektronenhirn im Tessin

«Ein äusserst komplizierter Rechtsfall kommt auf uns zu»

MANNO (bk) - Erste Hindernisse verzögern den rasanten Einzug des Supercomputers ins Tessin: Die Anmietung eines fixfertigen Rechenzentrums in Manno klappt doch nicht so reibungslos, wie Bund und Eidgenössische Technische Hochschule (ETH) sich das vorgestellt haben. Dazwischengeschaltet haben sich mit einer Klage die ehemaligen Besitzer des Geländes.

Ihnen war das 140 000 Quadratmeter grosse Grundstück in Manno in den siebziger Jahren enteignet worden. Es war von den SBB als Standort des geplanten Güterbahnhofs Lugano-Vedeggio ausersehen. Doch für den Bau des "Cargo Domicilio" war dann nur ein Teil des Geländes notwendig. Die Bundesbahnen beschlossen deshalb, den nicht benötigten Rest zu vermieten. Und zwar für jährlich zehn Franken pro Quadratmeter und mit eine Mietvertragsdauer von 60 Jahren.

Unter den Mietern figuriert auch der Luganeser Bauunternehmer Silvio Tarchini. Er will auf dem Grundstück ein 16 500 Quadratmeter grosses Industriezentrum bauen und hat einen Teil davon - 2000 Quadratmeter - dem Amt für Bundesbauten zur Miete angeboten. Die Offerte kam gerade recht, denn das Amt hielt nach einem Standort für ihr nationales Zentrum für Hochleistungsrechner Ausschau. Ausserdem hatte Architekt Tarchini gleich auch noch auf eigenes Risiko ein Projekt für einen Pavillon zur Installierung des Elektronenhirns und der dazugehörenden Büros ausgearbeitet (TZ vom 20. Mai).

Baubeginn trotz unklarer Besitzverhältnisse?

Dass die SBB ihr ehemaliges Grundstück nun einfach gewinnbringend vermieten, stösst den früheren Besitzern des Geländes sauer auf. Sie reichten Klage beim Gericht ein. Denn gemäss Artikel 104 des Enteignungsgesetzes müssen enteignete Besitzer informiert werden, wenn ihr ehemaliger Boden plötzlich einem anderen als dem Enteignungszweck zugeführt wird. Dies haben die SBB jedoch unterlassen.

Der Bellinzoneser Rechtsanwalt Tuto Rossi, der die Bundesbahnen in dieser Angelegenheit vertritt, hat eine harte Nuss zu knacken. Die rechtliche Lage, mit der er es hier zu tun hat, ist äusserst kompliziert. Und sie könnte noch verzwickter werden: «Sollte das Gericht zugunsten der Kläger entscheiden, hätten wir es mit einem der schwierigsten Rechtsfälle überhaupt zu tun», orakelt der *avvocato*. Das Grundstück könnte dann an die früheren Besitzer zurückgehen - allerdings um ein privates Industriezentrum und einen bundeseigenen Supercomputer angereichert.

Derweil hat das von Architekt Tarchini geplante Bauvorhaben - vorläufig noch ungestört - die ersten Hürden genommen. Das *Municipio* von Manno hat dem Unternehmer bereits eine Baubewilligung erteilt. Ein paar Auflagen sind dem Architekten zwar aufgebrummt worden. Aber die sind eher nebensächlicher Natur - wie beispielsweise die Anordnung, dass ein Teil der 400 im Freien vorgesehenen Parkplätze unterirdisch angelegt werden soll. Die abgeänderten Baupläne werden in ein paar Tagen an die Gemeinde zurückgehen. Dann könnte mit den Bauarbeiten begonnen werden. In der Zwickmühle sitzen derweil der Bund und die ETH, die angesichts der unklaren Besitzverhältnisse nicht so recht wissen, wohin mit ihrem Computer.

Fig. 1: Tessiner Zeitung, 3 June 1989, «First hurdles for the planned supercomputer in Ticino» (CSCS Archive, 1989 Manno – Nationales Hochleistungsrechenzentrum, Press articles, Tessiner Zeitung, 3/6/1989).

After learning about this problematic situation, on 14 June 1989, the Swiss federal government warned that a final decision on the location was not to be taken too quickly, since the project was extremely important on a national level and the consequences of a bad decision would seriously impact its completion. Along with the problematic legal disputes, it was pointed out that «[neither] the location in a highly industrial zone nor the buildings se-

will auf dem Grundstück ein 16 500 Quadratmeter grosses Industriezentrum bauen und hat einen Teil davon – 2000 Quadratmeter – dem Amt für Bundesbauten zur Miete angeboten. [...] Dass die SBB ihr ehemaliges Grundstück nun einfach gewinnbringend vermieten, stösst den früheren Besitzern des Geländes sauer auf. Sie reichten Klage beim Gericht ein.»

lected seem to reflect the importance of the project».[45] In addition, «[it] would have been better to house the computer in a specially designed and constructed building», and to actively involve the Canton of Ticino with «a regular call for proposals»,[46] also considering that «tempting offers from other entrepreneurs in the area» had been rejected without being considered.[47]

Moving the deadline for the definite decision on the location to 30 March 1990,[48] and the deadline for the final installation of the center to 30 June 1991,[49] the working group started to look for other possible locations in case the dispute between the former owner of the space in Manno and the SBB was not quickly resolved.[50] On 24 October 1989, a public request for proposals was published by the Canton of Ticino, and in one month 36 different proposals were submitted.[51] The proposals included Tarchini's original site in Manno (the dispute over which was now resolved), as well as one from a pair of brothers, Giuliano and Attilio Bignasca, in the nearby town of Bioggio (which had previously been considered as an option but was discarded due to non-compliance) (Fig. 2). These two projects appeared to be

45 Ibid., Parliament, Interpellation of Mr. Carobbio, 14/6/1989. Original text in German: «Weder der Standort in einer ausgesprochen industriellen Zone noch die ausgewählten Gebäude scheinen der Bedeutung des Projektes Rechnung zu tragen. [...] Besser wäre, den Computer in einem eigens dafür konzipierten und errichteten Gebäude unterzubringen.»

46 CSCS Archive, 1988–1991 CSCS – Zeitungsartikel Entwicklung Infrastruktur Manno, Gazzetta Ticinese, 22/8/1989. Original text in Italian: «un regolare bando sul Foglio ufficiale».

47 Ibid., Giornale del Popolo, 12/9/1989. Original text in Italian: «rifiuti di allettanti offerte di altri imprenditori della zona che permetterebbero di installare il Centro di calcolo ad un tiro di schioppo da Manno».

48 ETHZ Archives, Manno_Akz2000-01_Auswahl, 1989–1991_POL HLR-91_12, Minutes meeting HLR-91-Infrastruktur, 18/9/1989, p. 2.

49 Ibid, p. 4.

50 CSCS Archive, Project HLR-91, Realization phase, Swiss Federal Chancellery on CSCS, 2/10/1989.

51 ETHZ Archives, Manno_Akz2000-01_Auswahl, 1989–1991_POL HLR-91_8, p. 3. Press release on the continuation of work on CSCS location in Ticino, 1/12/1989.

the only ones «able to meet the 1991 deadline»,[52] but this two-horse race led to a local dispute over the location of CSCS. In particular, the Bignascas reproached the government for not previously publishing a public request for proposals, but rather favoring a near ready-made project for time reasons to the detriment of normal competition. The Bignascas wanted the public to be aware that Tarchini's initial offer and, consequently, the government's decision had not followed regular procedures. In a local newspaper, Giuliano Bignasca now claimed to be very satisfied the government had finally decided to open the public request for proposals, arguing that «[i]n our country there is a competition to award the construction of a few meters of sewer. Why should the same not be done for a [...] supercomputing center?».[53]

In addition to the Tarchini–Bignasca affair, another protest against the initial choice of Tarchini's project came from a group of architects in Ticino. In a letter addressed to the Ticino State Council signed by prestigious figures, including Mario Botta, architects reclaimed their right to create a more «functional, decent, representative building of good quality» for CSCS and to safeguard the Ticinese cultural landscape.[54]

On 1 December 1989, Tarchini's proposal was pronounced «the only suitable project»[55] by the ETH Council. The project was accepted by the Fed-

52 CSCS Archive, 1988–1991 CSCS-Zeitungsartikel Entwicklung Infrastruktur Manno, Giornale del Popolo, 25/10/1989. Original text in Italian: «soltanto io [Bignasca] e Tarchini siamo in grado di rispettare la scadenza del 1991».

53 Ibid., 9/11/1989. Original text in Italian: «Al Poli dovrebbero essere pervenute una ventina di proposte, in seguito alla pubblicazione sul Foglio ufficiale e cioè da quando l'arrivo del Centro in Ticino ha preso una piega ‹pubblica›: una conclusione più che logica, ha aggiunto Bignasca. ‹*Nel nostro paese si fa un concorso per attribuire la costruzione di pochi metri di fognatura, perché non si doveva fare con un centro di calcolo da 20 milioni?*›»

54 CSCS Archive, 1989 Manno – Nationales Hochleistungsrechenzentrum, Architect. Letter from Belinelli and Galfetti to Ticino State Council, p. 2. Original text in Italian: «una struttura funzionale, dignitosa, rappresentativa: di buona qualità.»

55 ETHZ Archives, Manno_Akz2000-01_Auswahl, 1989–1991_POL HLR-91_8, pp. 4–5. Letter from Scaroni to Zehnder, 1/12/1989. Original text in German: «Besonders auffallend ist das Zitat von Herrn Tarchini für das Projekt in Manno, das einzige bewilligte Projekt». According to Caccia, the decision in favor of the Tarchini project was made during a meeting with federal councilor Flavio Cotti and finance minister Otto Stich.

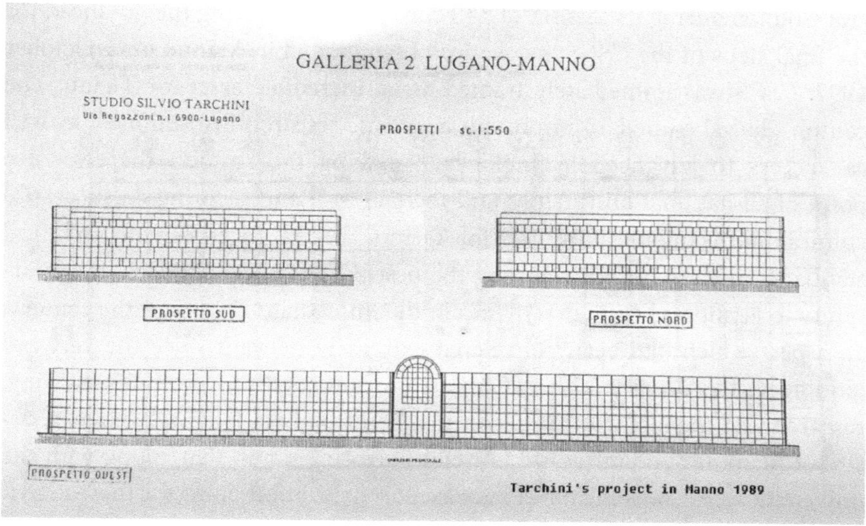

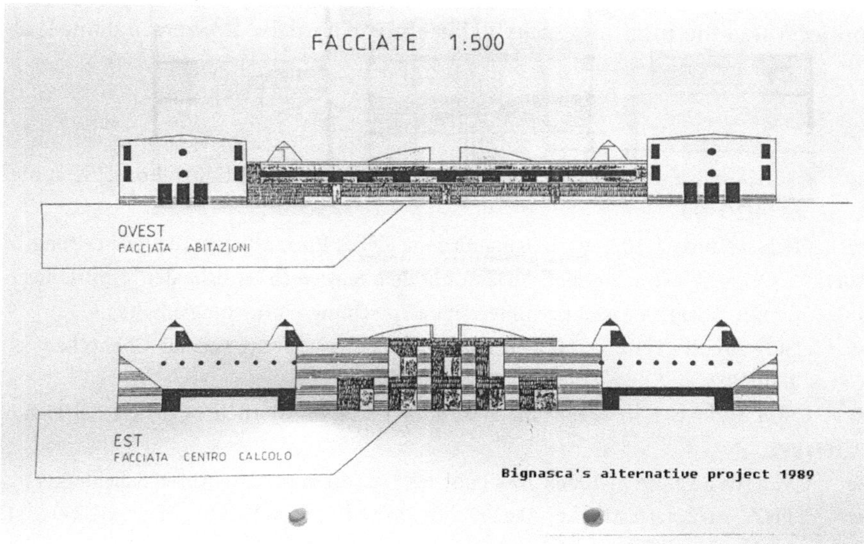

Fig. 2.1 and 2.2: The architectural plans of the competing buildings proposed by Tarchini and Bignasca in 1989 (CSCS Archive, 1989 Manno – Nationales Hochleistungsrechenzentrum, Tarchini, Stabile Galleria Manno 2; Bignasca, Alternative Project).

Stich would agree to a solution in Ticino only if it offered the most advantageous financial terms, which Tarchini's proposal did.

eral Council during its session of 14 February 1990,[56] giving the go-ahead for the final steps of the CSCS installation. Inaugurated in Manno on 1 October 1992, CSCS was immediately framed as an incredible asset for Ticino. The canton «ha[d] gained an important academic research institution»[57] as well as «access to advanced technology»,[58] making the region «the reference point of all national universities and the ETHs»[59] and securing «a successful cultural and economic future»[60] for Ticino. Indeed, together with technical standards, the criteria for selecting the best project also included the opportunity to «[initiate] further activities in the immediate vicinity of the center», an aspect which had been considered extremely relevant since the very beginning of the project. The CSCS was also intended «to promote teaching, research and technology transfer in the third national language» area, a desire of many for several years, as well as creating a hub that, along with the university sector, was meant «to act as a crystallization point for these activities», as well as «a number of projects» in higher education.[61] Among these projects was the local university (USI Università della Svizzera italiana), al-

56 CSCS Archive, Project HLR-91, Realization phase, Federal Council on HSC center location, 14/2/1990.

57 CSCS Archive, 1.10.1992 – Inaugurazione CSCS, Press articles, Tessiner Zeitung, 5/10/1992. Original text in German: «Denn mit dem Superrechner habe der Nicht-Universitäts-Kanton Tessin eine wichtige universitäre Forschungseinrichtung erhalten.»

58 CSCS Archive, Project HLR-91, Realization phase, Swiss Federal Chancellery on CSCS, 2/10/1989, p. 1.

59 CSCS Archive, 1.10.1992 – Inaugurazione CSCS, Press articles, Corriere del Ticino, 2/10/1992.

60 CSCS Archive, 1992–2000 Cross-cuts, 1994, p. 2, Editorial by Alfred Scheidegger.

61 ETHZ Archives, Manno_Akz2000-01_Auswahl, 1989–1991_POL HLR-91_11, Analysis of location options, 12/10/1989, p. 6. Original text in German: «Als zusätzlicher Aspekt für die Gesamtbeurteilung des Projektes gilt die Möglichkeit, weitere Aktivitäten in der unmittelbaren Umgebung des Zentrums anzusiedeln [...]. Dadurch wird die Realisierung des politischen Vorhabens Lehre, Forschung und Technologietransfer in der dritten Landessprache zu fördern, erst möglich. Neben dem universitären Bereich soll das HLR-Zentrum als Kristallisationspunkt für diese Aktivitäten wirken. Auf kantonaler, eidgenössischer und europäischer Ebene zeichnen sich bereits einige Projekte ab, die für eine Umsiedlung in die Nähe des Zentrums in Frage kommen.»

ready considered as a possible partner before its establishment in 1996 and the creation of its Faculty of Informatics in 2004.

Several of these projects actually saw the light of day (for example, USI), although the resistance of numerous actors, especially in Ticino and its State Council, seriously delayed or prevented cooperation between CSCS and other emerging schemes. According to Fulvio Caccia, the latter included «the concentration of scientific activities already present in Ticino (the Dalle Molle Institute for Artificial Intelligence, the Institute for Research in Biomedicine, and the Institute of Earth Sciences), or others underway, such as the IT and electronics sections of the University of Applied Sciences and Arts of Southern Switzerland (SUPSI)».[62]

If Tarchini was the winner of this local dispute, probably the most persistent local narrative around CSCS concerns the losers – the Bignasca brothers. According to them, their exclusion from the competition was an abuse of power by national stakeholders (the federal government, ETHZ, and SBB in particular). A few months later, the Bignascas decided to found the Lega dei Ticinesi, a regional movement intended to bring greater weight in Swiss decision-making processes for the Italian-speaking part of the country. Historically, it is uncertain whether the trigger for this development really was the «CSCS affair», but this narrative endures in Ticino.[63]

A bridge to «the highly industrialized north of Italy»: internationalization and regionalization

CSCS was not only envisioned as a national strategic asset but was also planned as a bridge for international connections. Notably, Ticino was considered a strategic location for connecting Swiss supercomputers (and Swiss research) to the rest of Europe, especially Northern Italy. This was a longstanding goal that emerged before the creation of the center; it was relevant to the decision-making process; and it persisted at least for the first decade of

62 Caccia, personal communication.

63 For example, in 2016 *La Regione*, one of the most important newspapers in Ticino, celebrated the 25th birthday of CSCS with an article reporting this story (see La Regione, 20/10/2016).

CSCS. But it was also in tune with a more general turn in Swiss and European politics in the 1980s and 1990s. In his *History of Switzerland*, Jakob Tanner states that the politics of the country embraced this European idea of regionalization and that Ticino was framed as one of the seven Swiss greater regions (*Grossregionen*), with the task of connecting itself to Milan and of respecting Swiss «federal sensitivities» (*föderalistische Empfindlichkeiten*).[64] The history and politics of CSCS are integral to the broader political economies and ideas that arose at that point in history.

A 1989 document compares the ten best locations of the 36 originally proposed. Among other reasons, Manno seemed to be the best option to establish a bridge with Europe, as evaluators from Zurich noted:

> ERCOFTAC. First contacts with EPFL show that the ERCOFTAC coordination group could be advantageously accommodated in Ticino. Synergies with SWITCH for the coordination, planning, and implementation of Europe-wide high-speed communication networks are conceivable. The chairman of ERCOFTAC in Brussels supports this idea.[65]

ERCOFTAC is the European Research Community on Flow, Turbulence and Combustion. It is a mixed public-private network of international collaboration and exchange between industry and research, and it promotes research centers for collaboration, simulation, and applied research. The main goal was to link CSCS with other research centers, to make it one of ERCOFTAC's nodes, and more generally to link Swiss research centers to European ones. In other words, placing the supercomputer in Manno meant opening a door to Europe.

On 2 October 1989, the Federal Council confirmed that locating the center in Manno would create an «important link to similar areas in north-

64 Jakob Tanner, Geschichte der Schweiz im 20. Jahrhundert, Munich 2015, pp. 481 and 1982.

65 ETHZ Archives, Manno_Akz2000-01_Auswahl, 1989-1991_POL HLR-91_11, Analysis of location options, 12/10/1989, pp. 6-7. Original text in German: «ERCOFTAC. Erste Kontakte mit EPFL zeigen, dass die Koordinationsgruppe des ERCOFTAC mit Vorteil im Tessin untergebracht werden könnte. Synergien mit SWITCH für die Koordination, Planung und Realisation von europaweiten schnellen Kommunikationsnetzen sind vorstellbar. Der Chairman von ERCOFTAC in Brüssel unterstützt diese Idee.»

ern Italy».[66] This is one of the first instances in which Italy is mentioned. Starting from the late 1980s, linking Switzerland and Swiss academic and scientific institutions to Italian ones would be obsessively repeated by different stakeholders.

As stated in a 1990 ETHZ press release:

> ETH Zurich would like to establish the national computer center in the Italian-language area in order to improve Swiss university relations with the area of the third national language and also to create a bridge to scientific partners in the highly industrialized north of Italy [...] which will also be politically significant for technology policy across borders.[67]

Note how improving «university relations» with Ticino was essentially a vague goal for ETHZ, whereas establishing links with industrialized partners in Northern Italy was considered «politically significant». This idea underpinned part of a strategy implemented by ETHZ since the second half of the 19th century, the so-called *Annexanstalten.* Research institutes such as EMPA (Eidgenössische Materialprüfungsanstalt) in Dübendorf and, later, the Paul Scherrer Institute in Würenlingen functioned as *annexes* to ETHZ, exploiting industrial interests and research with local partners and leaving academic and theoretical research to the polytechnic.[68]

66 CSCS Archive, Project HLR-91, Realization phase, Swiss Federal Chancellery on CSCS, 2/10/1989, p. 2.

67 ETHZ Archives, Manno_Akz2000-01_Auswahl, 1989-1991_POL HLR-91_3, Press release after the ETH Board meeting on 26-27/06/1990. Original in German: «Da alle Hochschulen der Schweiz im deutschen und im französischen Sprachgebiet liegen, möchte die ETH Zürich das nationale Rechenzentrum im italienischen Sprachgebiet errichten, um so die schweizerischen Hochschulbeziehungen zum Gebiet der dritten Landessprache zu verbessern und auch um eine Brücke zu wissenschaftlichen Partnern im hochindustrialisierten Norditalien zu schaffen [...] welcher auch für die Technologiepolitik über die Grenzen politisch bedeutungsvoll sein wird.»

68 On the ETHZ annexes policy, see David Gugerli, Patrick Kupper, et al., Transforming the Future. ETH Zurich and the Construction of Modern Switzerland 1855-2005, Zurich, 2010.

In September 1992, Giuseppe Buffi himself addressed a letter to the local department of education that outlined similar arguments but a broader strategic plan:

> The goal of this Swiss scientific computing center is to radiate the strength of its message and to find synergies with the places that host it, here, and perhaps as far as nearby Lombardy. [...] This represents a new paradigm for Ticino. It is the Ticino that many Ticinesi aspire to. [...] I really don't think we will stop here. Not out of self-serving ambition, but to participate with equal commitment, rights, and dignity in the progress of civil society with an eye to European horizons.[69]

This is another way of framing the same narrative from the vantage point of the local stakeholders. The about-to-be opened supercomputing center could usher in a new era for Ticino, a region often underestimated by the rest of Switzerland and by Italy itself. This was the origin of a new paradigm, according to Buffi, also because thanks to CSCS Ticino could bridge the rest of Switzerland to Italy and to a broader European perspective.[70] Buffi was right. CSCS proved to be a real turning point in the scientific and academic emergence of Ticino and in its opening up to the international scientific community. In 1986, for example, a referendum on the establishment of the university (Centro universitario della Svizzera italiana) was rejected by citizens. But the same politicians who lobbied in favor of CSCS, such as Buffi and Cotti, in the 1990s made the university a reality. Together, CSCS and USI Università della Svizzera italiana would attract scientists and students from all over the world, helping to internationalize the region.

During the CSCS inaugural ceremony, several addresses touched on the strategic relevance of the regional dimension. According to Alfred Scheideg-

69 CSCS Archive, 1.10.1992 – Inaugurazione CSCS, For the press, Letter from Buffi to the Ticino department of education, 30/9/1992, pp. 11–12. Original in Italian: «L'ambizione di questo centro svizzero di calcolo scientifico è di riuscire a irradiare attorno a sé la forza del suo messaggio, e di trovare sinergie con i luoghi che lo ospitano, qui, e magari fin giù nella vicina Lombardia. [...] Qui siamo su un quadrato di Ticino nuovo. È il Ticino cui molti ticinesi aspirano. [...] Credo proprio che non ci fermeremo qui. Non per vuota ambizione, ma per partecipare, a parità di sforzi, di diritti e di dignità, al progresso della società civile con uno sguardo rivolto alle prospettive europee.»

70 Interview with Mauro Martinoni, 25/8/2020.

ger, CSCS's first director, «the supercomputer [was] not only important for Swiss research and the economy, but it [would] also play an important role in Switzerland's research policy, which aim[ed] at integrating the country within Europe».[71] Among the new activities to develop the center, cooperation with universities, research centers, private industry, and computer manufacturers in Europe were mentioned. Similarly, ETHZ president Jakob Nüesch underscored the fact that the center enabled «access to Lombardy and its considerable potential. In this way, CSCS can become a relevant point of contact in the Swiss scientific and industrial communication networks».[72]

Heinrich Ursprung, former president of ETHZ and the newly elected state secretary for science, spoke at length on the origins and history of supercomputing in Switzerland. He mentioned four main reasons for locating the center in Ticino instead of either with EPFL or ETHZ, as seemed «logical»: the lack of space at the ETH Zurich main campus; the willingness to keep a national computer separate from those in the institutes of technology; the desire to establish high-tech scientific activities in Ticino; and, finally, «the value of an active approach – a European regional policy – toward Northern Italy».[73]

The regional and supranational role of CSCS continued to be quoted by new CSCS directors over time without significant changes. At the annual press conference in 1996, Jean-Pierre Thierre declared that CSCS aimed to act as:

71 CSCS Archive, 1.10.1992 – Inaugurazione CSCS, Press articles, Tessiner Zeitung, 5/10/1992. Original in German: «dass der Hochleistungsrechner nicht nur für die Schweizer Forschung und Volkswirtschaft von Bedeutung sei, sondern auch eine wichtige Rolle in der forschungspolitischen Integration der Schweiz in Europa spielen werde».

72 Ibid., For the press, Nüesch's discourse for CSCS inauguration 1/10/1992, p. 4. Original in German: «Der Standort in der italienischen Schweiz auf einem zukunftsträchtigen Gebiet, sondern erschliesst unserem Lande den Zugang zur Lombardei mit ihrem beträchtlichen Potential. Das CSCS kann zu einem wichtigen Glied eines wissenschaftlichen und industriellen Beziehungsnetzes unseres Landes werden.»

73 ETHZ Archives, Box 12_ER-GS-Leh01: 1.5.21 f, R-GS-Leh01: 1–5–21e-f, pp. 88–90, Discourse of Heinrich Ursprung (representing the Confederation) at the opening of CSCS in Manno. Original in French: «l'intérêt d'une ouverture active – politique européenne des régions – sur l'Italie du Nord».

> promoter of specific projects that may interest Ticino and Lombardy. [...] [CSCS] intends to play a bridge role, acting as a catalyst, between the research of higher education institutions and industry, public administration, [and] the academic community as a whole».[74]

The same key terms were invoked («Lombardy», «bridge», «catalyst»), the same idea of providing connections between the Swiss and Italian research communities.

In 1999 Michele Parinello, the new director of CSCS and later professor at USI Università della Svizzera italiana, delivered a speech in which he proposed a change of strategy for the Swiss supercomputing center, a shift from «a service center to a science center». The aim was to make CSCS a world-leading research center for computer-assisted simulations of physical, chemical, and biochemical processes. What did not change, yet again, was the focus on the research centers of Northern Italy, especially Lombardy: «CSCS opens up: Links with scientists and institutions in Switzerland and abroad are to be strengthened, particularly the collaborations with the Università della Svizzera italiana and Politecnico di Milano».[75]

Opening to Europe, and Northern Italy in particular, was one of the enduring discourses that emerged before, during, and after the founding of CSCS. Between the late 1980s and early 1990s, it was considered one of the justifications for placing the center in southern Switzerland and was part of the European policy of regionalization embraced by the country. Linking to Italy meant linking to Europe. The Italian border was so close to Manno and then to Lugano (about twenty kilometers away), that this link looked «natural». During the 1990s, referring to the link to Italy and Europe justified the existence of the center itself, which in its first years was beset by problems, including unstable management.

74 ETHZ Archives, Biographisches Dossier ETH/CSCS, Corriere del Ticino, 15/6/1996. Original in Italian: «promotore di progetti specifici che possano interessare Ticino e Lombardia [...] intende svolgere un ruolo di ponte, agendo come catalizzatore, tra la ricerca degli istituti di insegnamento superiori e l'industria, la pubblica amministrazione, la comunità accademica nel suo insieme.»

75 CSCS Archive, Annual Reports, 1999–2000, p. 3.

Despite these declarations, at least in the 1990s, research projects and industrial connections with Italian companies were lacking, and Italian partners were not privileged over other partners.[76] On the contrary, probably the strongest international link was with Japan and with a private company, NEC. Indeed, the NEC SC-3/22 was the first supercomputer installed at CSCS, and from 1993 CSCS signed an agreement with NEC to allow the Japanese company to «set up a research center at CSCS in order to develop software for supercomputers».[77] This collaboration, which brought Japanese engineers from NEC to Manno for collaboration «in the field of research and development»,[78] lasted for three years (Fig. 3). It was stopped in 1996.[79]

The multiple failures of regional and international collaborations in the early 1990s do not mean that this narrative was weak, specious, or even false (we do not have enough evidence to make such a claim). It was one of the narratives that helped different stakeholders to frame a simple and probably successful argument: Switzerland needs a supercomputer. Ticino is the right place to host it (also because it is close to Italy and thus to Europe). Consequently, the supercomputer will attract researchers and expand collaborations in the region and worldwide. In short, thanks to its supercomputer, Switzerland will be closer to Europe or at least on the map of regional and global collaborations.

---

76 ETHZ Archives, Biographisches Dossier ETH/CSCS, Neue Zürcher Zeitung, 20/5/1995. This article does indeed list companies that used the supercomputing center in Switzerland. Only one Italian company is listed (Agusta compagnia aeronautica italiana), which went bankrupt a few years later. The other companies mentioned are ABB, Ciba, Convex, and the supercomputer companies Cray and NEC.

77 Ibid. Original in German: «Die japanische Firma hat im CSCS eine Forschungsstelle zur Entwicklung von Software für Supercomputer eingerichtet.»

78 ETHZ Archives, Box 12_ER-GS-Leh01: 1.5.21f, R-GS-Leh01: 1-5-21e-f, p. 71, Invitation letter from Scheidegger to Grin, 23/2/1993. Original text in German: «Zusammenarbeitsvertrag zwischen dem Centro Svizzero di Calcolo Scientifico (CSCS) und der Firma NEC im Bereich Forschung und Entwicklung».

79 Ibid., p. 41, Letter from Grin to Faranak Grange on the end of the CSCS/NEC collaboration, 4/11/1996. Original in French: «[J]e vous donne quelques informations sur la collaboration à ce sujet, maintenant terminée, entre NEC et le Centre suisse de calcul scientifique (CSCS), à Manno (Tessin).»

Fig. 3: Group photo of the fifth NEC-ETHZ joint workshop, 11–12 May 1995 (ETHZ Archives, Akten des Vizepräsidenten Forschung der ETH Zürich, CSCS – NEC – Manno IV).

Conclusions: CSCS and the many faces of digital federalism

The birth of CSCS provides a series of compelling anecdotes and sources that relate directly or indirectly to the digital federalism concept. These sources show the complexity that the creation of an innovative scientific institution implies at the economic, political, and technological level. The history of CSCS also highlights some of the key issues that a country like Switzerland had to face during a technological transition in order to follow its federalist model. The role of CSCS in the digitization of Switzerland was, in fact, not only to provide Swiss people with a technological innovation. Rather, CSCS represented the attempt to address the transition to the information age in accordance with the principles of federalism and to maximize the potential of the Italian-speaking part of the country, which demanded to take part in national and global technoscientific advancements.

In particular, three key concepts that emerged from our research, and that are strictly linked to the preceding sections, perfectly summarize the symbolic and historical relevance of CSCS for Swiss digital federalism. These concepts embed both the discourses surrounding the birth of the center and federalism as a cultural, political and economic template.

First, CSCS was often compared to a «bridge» that would connect Ticino with the rest of Switzerland, and Switzerland to Northern Italy and thus to Europe (and maybe the world). This center would act as a national, intercantonal, regional, and international node, respectively, bringing Ticino and the other cantons closer together but also linking it with Swiss scientific institutions and Italian companies. This bridge would thus cross the local, national, and regional and international boundaries, becoming a strategic infrastructure for a techno-nationalist process as well as for the participation and involvement of Switzerland in European and international markets. Although some of these goals took time to accomplish, the idea of CSCS as a connecting bridge encouraged and legitimized the realization of the center in Ticino. The result was to provide the region, and Switzerland at large, with a strategic asset for its digital future.

A second key concept that emerged from our analysis is «decentralization». One aim of CSCS was to decentralize Switzerland's digital resources and assets. As political scientists and historians have aptly shown, decentralization is a key element both of Swiss federal policies and the country's economic choices.[80] Historically, decentralization has allowed cantons and local authorities to maximize and exploit their specific resources, responding to citizens' needs that local governments know much better than the federal government does. In our case, decentralization entailed the decision to create a new «center» in Ticino. At the same time, this decentralizing and re-centralizing effort also strengthened the established federal center in Zurich, especially at ETHZ. The tension between centralization and decentralization

80 Sean Mueller, Adrian Vatter, Federalism and Decentralisation in Switzerland, in: Ferdinand Karlhofer, Günther Pallaver (eds.), Federal Power-Sharing in Europe, Baden-Baden 2017, pp. 39–64.

was clothed in irony. As contemporaries from Zurich observed, during its early days CSCS was stigmatized by the use of ambivalent nicknames, such as «a flower» or «a cathedral» in the desert. However, this choice did provide Ticino with a strategic institution for its growth. It also spurred local actors to request greater involvement in the founding process. The internal political conflicts that emerged during the dispute about location reflect a side effect of the digital federal strategy: centralization of political power in the hands of the federal government and ETHZ. This unbalanced distribution of power and responsibility for such a big project is quite normal in federal countries. Another key principle of federalism is subsidiarity, according to which processes that cannot be managed at the local level – for example, owing to economic or structural constraints – should be delegated to higher authorities and actors such as the federal government. It is clear that the creation of an expensive and complex institution such as CSCS could not be accomplished solely by the Ticinese actors. The federal government and ETHZ were essential to the fulfillment of this project. Furthermore, the tension between centralized control and local actors who asked to be involved in the design and construction phases of the building contributed, as we have said, to the formation of the Lega dei Ticinesi.

This juxtaposition between decentralization and centralization can be also read through the lens of political economy and sociotechnical analysis of networking. As media and technology historians have shown, even if network architectures like the Internet and the web have been depicted as decentralized infrastructures, the economic, technological, and political control over them often operates at a central level.[81]

Tensions aside, CSCS was also an act of «solidarity» with Ticino. Solidarity is a key principle in any federal association as it entails the faith (from the Latin *fede*) of those allies, whether states or cantons, who tend to support each other in order to strengthen (to make more *solid*) the cohesion and

81 See, for example, the work by Vincent Mosco, The Digital Sublime: Myth, Power, and Cyberspace, New York 2005. On the tension between centralization and decentralization in network histories, see Francesca Musiani and Valérie Schafer, Le modèle internet en question (années 1970–2010), in: Flux, 2011, 3(85–86): 62–71; Paolo Bory, The Internet Myth: University of Westminster Press, London 2020, pp. 7–30.

development of the entire nation. The history of CSCS tells us that this act of solidarity took some decades to bear fruit and involved many institutions that had to find a coordinated strategy to achieve such an important milestone toward digital federalism. ETHZ, the federal government, the ETH Council, the Ticino parliament, SBB, professional associations, and opposition parties – all these actors were involved and contributed, for better or worse, to the collective effort to build the center. Although the first decades of this history were plagued with political and economic issues, CSCS eventually became a key reference within the Swiss and European scientific communities. In this regard, further research is essential to understand how the center «survived» its crisis during the 1990s and why and how Swiss actors have maintained their trust in this strategic asset.

Overall, the sources analyzed in this paper show what a complex and compelling concept digital federalism is. This idea is rooted in the mutual trust and cooperation among political and economic actors as well as in the delicate handling of the tensions between local, regional, and national forces. The history of CSCS is exemplary in this sense.

The many facets of digital federalism emerging from this narrative represent the stratification and multiplicity of the Swiss cultural, economic, and political environment. Most importantly, notwithstanding the weak points, these facets reveal the peculiar strength of the Swiss federal system, which contributed to the founding of a key institution for the digital future of the country.

Author Biographies

Gabriele Balbi is associate professor in media studies at the Institute of Media and Journalism, USI Università della Svizzera italiana, Switzerland. He chairs the ECREA Communication History section, and his recent research deals with the history of media and digital technologies. His most recent book (co-edited with Andreas Fickers) is *History of the International Telecommunication Union (ITU). Transnational Techno-Diplomacy from the Telegraph to the Internet* (De Gruyter, 2020). Address: IMeG – USI, Via Giuseppe Buffi 13, CH-6900 Lugano, email: gabriele.balbi@usi.ch

Paolo Bory is a postdoctoral fellow at the Department of Design at Politecnico di Milano, Italy. His research focuses on the histories and imaginaries of digital technologies such as the Internet, AI, and supercomputing. Paolo has published peer-reviewed articles on these topics in international journals such as *New Media & Society*, *Convergence*, and *Critical Studies in Media and Communication*. He is the author of *The Internet Myth: From the Internet Imaginary to Network Ideologies* (University of Westminster Press, 2020). Address: Dipartimento di Design, Via Durando 10, 20158 Milano IT, email: paolo.bory@polimi.it

David Gugerli, born in 1961, is professor of history of technology at ETH Zurich. His research is focused on historical changes in computational reality. Most recently he published *Wie die Welt in den Computer kam. Zur Entstehung digitaler Wirklichkeit* (S. Fischer Verlag, 2018) and David Gugerli and Ricky Wichum, *Simulation for All. The Politics of Supercomputing in Stuttgart* (Chronos Verlag, 2021). Address: Clausiusstrasse 59, RZ F11, CH-8092 Zurich, email: gugerli@ethz.ch.

Ely Lüthi is a PhD student and teaching assistant at the Institute of Media and Journalism at USI Università della Svizzera italiana, Switzerland. Her PhD research focuses on the history of Swiss digitization, Swiss political economy of communication and digitization policies. Her research interests also include (digital) media history, media policy and management, company cultures, and radio and television broadcasting. She has also been a research assistant for the project «Digital Federalism. The early history of

SWITCH and CSCS Manno (1985–1995)», funded by the Swiss National Science Foundation and jointly conducted with ETHZ. Address: IMeG – USI, Via Giuseppe Buffi 13, CH-6900 Lugano, email: ely.luethi@usi.ch

Moritz Mähr is a research assistant in the Chair of the History of Technology at ETH Zurich and at the Collegium Helveticum. He studied history and philosophy of knowledge, computer science, and banking and finance in Zurich and Berlin. He is currently working on a dissertation on the automation of migration authorities in Switzerland from 1964 to 1984. The study is part of the «Trading zones» project funded by the Swiss National Science Foundation. His research interests include computer history, migration studies, and digital history. Address: Institute of History, Chair for the History of Technology, ETH Zurich, Clausiusstrasse 59 (RZ F16), CH-8092 Zurich; email: mmaehr@ethz.ch

Nick Schwery has been a PhD candidate in the Chair for History of Technology at ETH Zurich since 2018. He studied social and communication sciences at the University of Lucerne from 2008 to 2011 and history and philosophy of knowledge at ETH Zurich from 2014 to 2017. His dissertation examines the genesis of the Hermes project management method and the structuring of the space between the computer and the federal administration between 1960 and 1976. Address: ETH Zurich, Institut für Geschichte, RZ F16, Clausiusstrasse 59, 8092 Zurich, email: nick.schwery@history.gess.ethz.ch

Laura Skouvig is associate professor at the Department of Communication, University of Copenhagen. Her research field is information history and early modern Denmark in the period from 1784 to 1820. The focus of her research, and her publications, is information cultures, information networks, and implications for surveillance and communication in late absolutist Denmark. Address: University of Copenhagen, South Campus, Karen Blixens Plads 8, 2300 Copenhagen S, Denmark, email: laura.skouvig@hum.ku.dk

Ricky Wichum, born 1982, is a research assistant at the Institute of History (History of Technology), at ETH Zurich. His recent work examines the use of information systems in the Swiss federal administration (Swiss National Science Foundation project «Trading Zones. Computers and Swiss Federal

Administration, 1960–1990»). Selected publications include *Biometrie. Zur Soziologie der Identifikation*, Wilhelm Fink-Verlag, Munich 2017, and David Gugerli and Ricky Wichum, *Simulation for All. The Politics of Supercomputing in Stuttgart* (Chronos Verlag, 2021). Address: Clausiusstrasse 59, RZ F15, CH-8092 Zurich, email: rwichum@ethz.ch

Daniela Zetti is a historian of technology and a postdoctoral researcher at the University of Lübeck, Germany. She teaches interdisciplinary courses in the medical, computer, and engineering sciences. Since April 2020 she has been vice head of the Ethical Innovation Hub at the University of Lübeck. Her topics of research are technology, innovation, and data sovereignty. She recently published, together with Christine Hämmerling, *Das dokumentierte Ich. Wissen in Verhandlung* (Chronos Verlag, 2018). Address: University of Lübeck, Institute for the History of Medicine and Science Studies, Königstrasse 42, D-23552 Lübeck, Email: daniela.zetti@uni-luebeck.de